THEORY OF SAMPLE SURVEYS

THEORY OF SAMPLE SURVEYS

Arjun K Gupta

Bowling Green State University, USA

D G Kabe

St Mary's University, Canada

World Scientific

NEW JERSEY • LONDON • SINGAPORE • BEIJING • SHANGHAI • HONG KONG • TAIPEI • CHENNAI

Published by

World Scientific Publishing Co. Pte. Ltd.
5 Toh Tuck Link, Singapore 596224
USA office: 27 Warren Street, Suite 401-402, Hackensack, NJ 07601
UK office: 57 Shelton Street, Covent Garden, London WC2H 9HE

British Library Cataloguing-in-Publication Data
A catalogue record for this book is available from the British Library.

THEORY OF SAMPLE SURVEYS

ISBN-13 978-981-4322-47-8
ISBN-10 981-4322-47-4

Printed in Singapore by World Scientific Printers.

To Arhaam

Preface

Once the first author was asked what is the most important branch of statistics? His answer then was, and still is, "It is Sample Surveys." For without sample surveys there is no data and without data there is no statistics. He has taught Sample Surveys in India as Lecturer in Statistics and also in Ghana (West Africa) as United Nations Consultant. The second author taught Sample Surveys in India as well as in Canada. This book is a growth of part of their lecture notes. It is theoretical in the sense that it gives mathematical proofs of the results in Sample Surveys. It is intended as a textbook for a one-semester course for undergraduate seniors or first-year graduate students. Basic knowledge of algebra, calculus, and statistical theory is required to master the techniques described here.

There are ten chapters in this book. Chapters 1, 2, and 3 deal with the simple random sampling, sampling with varying probabilities of selection, and stratified sampling, respectively. Chapters 5 and 6 describe methods of ratio estimation and regression estimation. The next three, Chapters 7, 8, and 9 present the results from cluster sampling, two-stage and three-stage sampling and double sampling, respectively. Finally non-sampling error is studied in Chapter 10. The presentation is systematic and rigorous. Every chapter is followed by a set of exercises which extend the concepts. A table of random numbers is given in the Appendix. The authors wish to acknowledge their appreciation to Professors C. R. Rao and Wei Ning for their help with the preparation of the manuscript. A bibliography and subject index are also provided.

The authors are thankful to many graduate students who read parts of the manuscript. Particular mention should be made of Deniz Akdemir, Ngoc Nguyen, Nan-cheng Su, and Wan-Ping. The editorial help given by Nisha and Patrick is worthy of specific note and is particularly well appreciated. Finally we would like to thank Cyndi Patterson, whose word processing skills made the preparation of this manuscript far less painful than it might have been. It has been a pleasure working with her.

A. K. Gupta
Bowling Green, USA

D. G. Kabe
Mississauga, Canada

Contents

Chapter 1

SIMPLE RANDOM SAMPLING

1.1 Introduction

For the purpose of the study of the sampling methods, it is assumed that the population consists of N well-defined units. A sample is only a finite portion of the population. In this book, we shall study (i) the different methods of drawing a sample from the population and (ii) how to obtain estimates of the population parameters for each method of sampling. We assume that N is finite.

A sample is taken by drawing units one after another from the population. If the unit drawn at the previous stage is replaced into the population before the next draw is made, the sampling is called sampling with replacement and if it is not, then the sampling is called sampling without replacement.

1.2 Simple Random Sampling without Replacement

If a sample is drawn, unit by unit, without replacement, such that there is equal probability of selection for every unit at each draw, then the sample is called saimple random sample without replacement and the sampling procedure is termed simple random sampling without replacement.

We assume that a simple random sample is drawn without replacement from a finite population. We adopt the following notations:

$N(n)$ = Population (sample) size

$\bar{Y}(\bar{y})$ = Population (sample) mean, defined by

$$\bar{Y} = \sum_{i=1}^{N} y_i/N, \quad \text{and} \quad \bar{y} = \sum_{i=1}^{n} y_i/n.$$

$S_y^2(s_y^2)$ = Population (sample) mean sum of squares, defined by

$$S_y^2(N-1)=\sum_{i=1}^{N}(y_i-\bar{Y})^2,\ \text{and}\ s_y^2(n-1)=\sum_{i=1}^{n}(y_i-\bar{y})^2.$$

f = sampling fraction, defined by n/N.

The following theorem gives an unbiased estimator of the population mean and its variance.

Theorem 1.2.1. *If a simple random sample is drawn without replacement from a population, then the following results hold:*

(i) $E(\bar{y})=\bar{Y}$
(ii) $Var(\bar{y})=\frac{1-f}{n}S_y^2$.

Proof. With the ith unit of the population, we associate a random variable a_i defined as follows:

$$\begin{aligned} a_i &= 1, \text{ if the } i^{th} \text{ unit occurs in the sample} \\ &= 0, \text{ if the } i^{th} \text{ unit does not occur in the sample} \\ &\quad (i=1,2,\ldots,N). \end{aligned}$$

Then,

$$\begin{aligned} E(a_i) &= 1\times \text{Probability that the } i^{th} \text{ unit is included in the sample} \\ &= \frac{n}{N},\quad i=1,2,\ldots,N. \end{aligned}$$

$$\begin{aligned} E(a_i^2) &= 1\times \text{Probability that the } i^{th} \text{ unit is included in the sample} \\ &= \frac{n}{N},\quad i=1,2,\ldots,N. \end{aligned}$$

$$\begin{aligned} E(a_ia_j) &= 1\times \text{Probability that the } i^{th} \text{ and } j^{th} \text{ units are included in the sample} \\ &= \frac{n(n-1)}{N(N-1)},\quad i\neq j=1,2,\ldots,N. \end{aligned}$$

From the above results, one easily obtains

$$\begin{aligned} Var(a_i) &= \frac{n(N-n)}{N^2}, i=1,2,\ldots,N \\ Cov(a_i,a_j) &= -\frac{n(N-n)}{N^2(N-1)}, i\neq j=1,2,\ldots,N. \end{aligned} \tag{1.2.1}$$

Now, we can write $\bar{y}$ as

$$\bar{y} = \frac{1}{n}\sum_{i=1}^{N} a_i y_i.$$

Then,

$$E(\bar{y}) = \frac{1}{n}\sum_{i=1}^{N} E(a_i) y_i = \bar{Y},$$

and

$$Var(\bar{y}) = \frac{1}{n^2} Var\left(\sum_{i=1}^{N} a_i y_i\right) = \frac{1}{n^2}\left[\sum_{i=1}^{N} Var(a_i) y_i^2 + \sum_{i \neq j}^{N} Cov(a_i, a_j) y_i y_j\right].$$

Substituting the values of $Var(a_i)$ and $Cov(a_i, a_j)$ as obtained in (1.2.1) in the above expression of $Var(\bar{y})$ and simplifying, we obtain

$$Var(\bar{y}) = \frac{1-f}{n} S_y^2. \tag{1.2.2}$$

Thus, the theorem is proved.

We shall now proceed to obtain an unbiased estimator of $Var(\bar{y})$.

Theorem 1.2.2. *If a simple random sample is drawn without replacement from a population, then*

$$E(s_y^2) = S_y^2.$$

Proof. We have

$$s_y^2 = \frac{1}{(n-1)}\left[\sum_{i=1}^{n} y_i^2 - n\bar{y}^2\right] = \frac{1}{(n-1)}\left[\sum_{i=1}^{N} a_i y_i^2 - n\bar{y}^2\right].$$

Hence, taking, expectation, we get

$$E(s_y^2) = \frac{1}{(n-1)}\left[\sum_{i=1}^{N} E(a_i) y_i^2 - n\{Var(\bar{y}) + \bar{Y}^2\}\right].$$

Substituting the values of $E(a_i)$ and $Var(\bar{y})$ in the above and simplifying, we get the required result.

Theorem 1.2.3. *If a simple random sample is drawn without replacement from a population, then an unbiased estimator of $Var(\bar{y})$ is given by*

$$Est.Var(\bar{y}) = \frac{1-f}{n} s_y^2.$$

We now give below some results which are useful in making statistical inference from the sample.

1. Standard Error of $\bar{y}$:

$$S.E.(\bar{y}) = \sqrt{Var(\bar{y})} = \sqrt{\frac{1-f}{n}S_y^2}. \quad (1.2.3)$$

2. Estimated Standard Error of $\bar{y}$:

$$Est.S.E.(\bar{y}) = \sqrt{\frac{1-f}{n}s_y^2}.$$

3. $100(1-\alpha)\%$ confidence interval for $\bar{Y}$:

$$\bar{y} \pm t_{\alpha/2,n-1} \cdot Est.S.E.(\bar{y}), \quad (1.2.4)$$

where $t_{\alpha/2,n-1}$ is the upper $100(\alpha/2)\%$ percent value of t distribution with $(n-1)$ d.f.

4. Relative Standard Error of $\bar{y}$:

$$Rel.S.E.(\bar{y}) = \frac{S.E.(\bar{y})}{\bar{Y}}, \quad (1.2.5)$$

5. Estimated Relative Standard Error of $\bar{y}$:

$$Est.Rel.S.E.(\bar{y}) = \frac{Est.S.E.(\bar{y})}{\bar{y}}. \quad (1.2.6)$$

Sometimes, we may be interested in estimating the population total $Y = N\bar{y}$ rather than its mean $\bar{Y}$. For this purpose, one can easily derive the following results from those of $\bar{y}$. An unbiased estimator of the population total Y is given by $\hat{Y} = N\bar{y}$ with variance

$$Var(N\bar{y}) = N^2\frac{(1-f)}{n}S_y^2. \quad (1.2.7)$$

The standard error of $\hat{Y}$ is given by

$$S.E.(\hat{Y}) = N \cdot \sqrt{\frac{1-f}{n}S_y^2}, \quad (1.2.8)$$

and its estimate by

$$Est.S.E.(\hat{Y}) = N \cdot \sqrt{\frac{1-f}{n}s_y^2}. \quad (1.2.9)$$

The $100(1-\alpha)\%$ confidence interval for Y is given by

$$\hat{Y} \pm t_{\alpha/2,n-1}Est.S.E.(\hat{Y}). \quad (1.2.10)$$

As a generalization of result (i) of Theorem 1.2.1, we give below Theorem 1.2.4, which will be useful later on.

Theorem 1.2.4. *Let $f(y)$ be a real-valued function defined on every unit of the population. If a simple random sample of size n is drawn without replacement from a population of size N, then*

$$E\left[\frac{1}{n}\sum_{i=1}^{n} f(y_i)\right] = \frac{1}{N}\sum_{i=1}^{N} f(y_i).$$

As an illustration of Theorem 1.2.4, we have

$$E\left[\frac{1}{n}\sum_{i=1}^{n} y_i^2\right] = \frac{1}{N}\sum_{i=1}^{N} y_i^2.$$

1.3 Simple Random Sampling with Replacement

If a sample is drawn, unit by unit with replacement, such that there is equal probability of selection for every unit at each draw, then the sample is called simple random sample with replacement, and the sampling procedure is called simple random sampling with replacement.

We assume the same notations as in Section 1.2. Next we prove the following theorem which provides the estimator of the population mean.

Theorem 1.3.1. *In SRSWR,*

(i) $E(\bar{y}) = \bar{Y}$
(ii) $Var(\bar{y}) = \frac{N-1}{nN} S_y^2$.

Proof. Let a random variable a_i, associated with the ith unit of the population be defined as the number of times the ith unit occurs in the sample, $i = 1, 2, \ldots, N$. Clearly, a_i assumes values $0, 1, 2, \ldots, n$. The joint distribution of $a_1, a_2, \ldots, a_N$ is the following multinomial distribution

$$P(a_1, a_2, \ldots, a_N) = \frac{n!}{\prod_{i=1}^{N} a_i!} \cdot \frac{1}{N^n}, \tag{1.3.1}$$

where $\sum_{i=1}^{N} a_i = n$. For this multinomial distribution, the following results are easily verified.

$$E(a_i) = \frac{n}{N}, \qquad Var(a_i) = \frac{n(N-1)}{N^2}, \quad i = 1, 2, \ldots, N. \tag{1.3.2}$$

$$Cov(a_i, a_j) = -\frac{n}{N^2}, \qquad i \neq j = 1, 2, \ldots, N. \tag{1.3.3}$$

Now, we can write $\bar{y}$ as

$$\bar{y} = \frac{1}{n}\sum_{i=1}^{N} a_i y_i.$$

Hence, taking expectation of $\bar{y}$ and substituting the value of $E(a_i)$ from (1.3.2), one easily obtains

$$E(\bar{y}) = \bar{Y}.$$

Further,

$$Var(\bar{y}) = \frac{1}{n^2}\left[\sum_{i=1}^{N} Var(a_i)y_i^2 + \sum_{i\neq j}^{N} Cov(a_i, a_j)y_i y_j\right]. \tag{1.3.4}$$

Substituting, the values of $Var(a_i)$ and $Cov(a_i, a_j)$ from (1.3.2) and (1.3.3) in (1.3.4), and simplifying we get

$$Var(\bar{y}) = \frac{N-1}{nN}S_y^2. \tag{1.3.5}$$

Theorem 1.3.2. *In SRSWR,*

$$E(s_y^2) = \frac{N-1}{N}S_y^2.$$

Proof. We have

$$(n-1)s_y^2 = \sum_{i=1}^{n} y_i^2 - n\bar{y}^2 = \sum_{i=1}^{N} a_i y_i^2 - n\bar{y}^2, \tag{1.3.6}$$

where a_i's are defined in Theorem 1.3.1 Taking expectation of (1.3.6), we obtain

$$\begin{aligned}(n-1)E(s_y^2) &= \sum_{i=1}^{N} E(a_i)y_i^2 - n\{Var(\bar{y}) + \bar{Y}^2\} \\ &= \frac{n}{N}\sum_{i=1}^{N} y_i^2 - n\cdot\frac{(N-1)}{nN}S_y^2 - n\bar{Y}^2 \\ &= \frac{(n-1)(N-1)}{N}S_y^2, \end{aligned} \tag{1.3.7}$$

from which the theorem follows.

Theorem 1.3.3. *An unbiased estimator of variance of $\bar{y}$, in SRSWR, is given by*

$$Est.Var(\bar{y}) = s_y^2/n.$$

The proof follows from Theorems 1.3.1 and 1.3.2.

Thus we have the following results.

$$S.E.(\bar{y}) = \sqrt{\frac{N-1}{nN} S_y^2} \tag{1.3.8}$$

$$Est.S.E.(\bar{y}) = \sqrt{\frac{s_y^2}{n}}. \tag{1.3.9}$$

The $100(1-\alpha)\%$ confidence interval for the population mean $\bar{Y}$ is given by

$$\bar{y} \pm t_{\alpha/2,n-1} Est.S.E.(\bar{y}). \tag{1.3.10}$$

As regards the estimation of the population total Y, we have the following results:

Unbiased estimate of the population total Y is given by

$$\hat{Y} = N\bar{y},$$

with variance

$$Var(\hat{Y}) = N^2 \cdot Var(\bar{y}) = \frac{N(N-1)}{n} S_y^2 \tag{1.3.11}$$

$$S.E.(\hat{Y}) = N \cdot S.E.(\bar{y}) = \sqrt{\frac{N(N-1)}{n} S_y^2} \tag{1.3.12}$$

$$Est.S.E.(\hat{Y}) = N \cdot \sqrt{\frac{s_y^2}{n}}. \tag{1.3.13}$$

The $100(1-\alpha)\%$ confidence interval for the population total Y is given by

$$\hat{Y} \pm t_{\alpha/2,n-1} Est.S.E.(\hat{Y}). \tag{1.3.14}$$

In SRSWR, the same unit may occur more than once. Hence, an alternative estimator for $\bar{Y}$ can be based on the distinct units in the sample. Suppose that in SRSWR of size n, $y_1', y_2', \ldots, y_d'$ denote the values of d distinct units ($d \leq n$). Then, the following estimator is suggested.

$$\bar{y}' = \frac{1}{d} \sum_{i=1}^{d} y_i'. \tag{1.3.15}$$

For obtaining the expected value and variance of (1.3.15), it may be noted that two stages of randomization are involved:

(i) d is a random variable assuming values $1, 2, \ldots, n$ and
(ii) selection of d distinct units from N units with equal probability without replacement. Hence

$$E(\bar{y}') = E_1 E_2(\bar{y}'|d) = E_1 E_2\left(\frac{1}{d}\sum_{i=1}^{d} y_i'|d\right) = E_1(\bar{Y}) = \bar{Y}, \quad (1.3.16)$$

which shows that $\bar{y}'$ is an unbiased estimator of $\bar{Y}$. Further,

$$\begin{aligned} Var(\bar{y}') &= V_1 E_2(\bar{y}'|d) + E_1 V_2(\bar{y}'|d) \\ &= V_1(\bar{Y}) + E_1\left\{\left(\frac{1}{d} - \frac{1}{N}\right) S_y^2\right\} \\ &= 0 + \left[E\left(\frac{1}{d}\right) - \frac{1}{N}\right] S_y^2 \\ &= \left[E\left(\frac{1}{d}\right) - \frac{1}{N}\right] S_y^2. \end{aligned} \quad (1.3.17)$$

One can easily verify that

$$E\left(\frac{1}{d}\right) = \frac{1^{n-1} + 2^{n-1} + \ldots + N^{n-1}}{N^n}. \quad (1.3.18)$$

An unbiased estimator of the $Var(\bar{y}')$ is given by

$$Est.Var(\bar{y}') = \left[\left(\frac{1}{d} - \frac{1}{N}\right) + \frac{N-1}{N^n - N}\right] s_d^2,$$

where $s_d^2 = 0$ for $d = 1$ and $s_d^2 = \sum_{i=1}^{d}(y_i' - \bar{y}')^2/(d-1)$, for $d \geq 2$.

1.4 Procedures for Selection of Simple Random Samples

Units in the population are serially numbered from 1 to N. The procedure of selecting units from a population of N units with equal probability makes use of Tables of Random Numbers. Several tables of random numbers are available. We mention the following ones:

(i) Tippett's (1927) random number tables consisting of 41,600 random digits grouped into 10,400 sets of four-digit random numbers;

(ii) Fisher and Yates (1943) table of random numbers consisting of 15,000 random digits arranged into 1,500 sets of ten-digit random numbers;

(iii) Kendall and Smith (1939) table of random numbers with 100,000 random digits grouped into 25,000 sets of four-digit random numbers;

(iv) Rand Corporation (1955) table of random numbers with 1,000,000 random digits grouped into 200,000 sets of five-digit random numbers; and

(v) Rao, Mathai, and Mitra (1966) table of random numbers with 20,000 random digits arranged into 5,000 sets of four-digit random numbers.

One page from (v) is reproduced in Appendix A for illustrating the use of random numbers in selecting a simple random sample.

We describe below the procedures for selecting a simple random sample with or without replacement. If we wish to select a simple random sample without replacement, and if some unit is repeated in the sample, then we reject one and select another, avoiding thus the repetition of units in the sample.

Procedure 1. Let N, the population size, be a r-digit number. Suppose we wish to draw a SRS of size n. We refer to a page of a Table of Random Numbers and consult a r-digit column of random numbers. From this column of r-digit random numbers, we select one by one r-digit random numbers which are less than or equal to N, till we get n random numbers. The units whose serial numbers are these selected n random numbers constitute the simple random sample of size n. If some unit is repeated in the sample and if our purpose is to select a SRSWOR, then the corresponding selected random number is rejected and another r-digit random number is selected.

As an illustration, let $N = 190$ and $n = 10$. Since N is a 3-digit number, we see 3-digit column of random numbers in the page given in Appendix A. We begin with the 3-digit column starting from the left. We select first ten 3-digit random numbers which are less than or equal to 190. We find that these numbers are

112, 059, 112, 116, 124
090, 037, 078, 092, 062.

Hence, units with the above serial numbers constitute the required simple random sample of size 10. Here, we note that unit with serial number 112 is repeated twice. If we want a SRSWOR, then the repeated number 112 is rejected, and one more 3-digit number is selected. In this illustration, the next number is 155. Hence, the sample consists of units with serial numbers

112, 059, 116, 124, 090
037, 078, 092, 062, 155.

We note here that in this procedure, sometimes more random numbers are rejected, since all r-digit numbers greater than N and $\overset{\longleftarrow\ r\ \longrightarrow}{0\ 0\ \ldots\ 0}$ are not considered at the time of selection. To overcome this difficulty, other selection procedures are suggested which are given below.

Procedure 2 (Remainder Approach). Let N, the population size, be a r-digit number and let N' be the highest r-digit multiple of N. In the page of a Table of Random Numbers, we see the r-digit column of random numbers. From this column, we select r-digit numbers, one by one, which are less than or equal to N', till we get n numbers. We divide these numbers by N and obtain remainders. The units whose serial numbers are these remainders are selected in the sample. If the remainder is zero, then the unit with serial number N is selected.

As an illustration, let $N = 190$, and $n = 10$. Then, $N' = 950$. Next, from the page of table of random numbers given in Appendix A, we find from the 3-digit column of random numbers from the left, the following ten numbers which are less than or equal to 950.

343, 613, 580, 629, 472
769, 349, 894, 112, 433.

These numbers are divided by 190 and the following remainders are obtained:

153, 43, 10, 59, 92
9, 159, 134, 112, 53.

Then, the units with above serial numbers are selected in the sample.

Procedure 3 (Quotient Approach). Let N, the population size be a r-digit number and let N' be the highest r-digit number. Let $N' = kN$. In the page of Table of Random Numbers, we consult the r-digit column of random numbers. From this column, we select r-digit random numbers, one by one, which are less than or equal to $N' - 1$ till n numbers are obtained. We divide these numbers by k and obtain the quotients. If a particular quotient is i, then the unit with serial number $(i + 1)$ is selected.

As an illustration, let $N = 190$, and $n = 10$. Then, $N' = 950$ and $k = 5$. From the 3-digit column of random numbers from the page of Table of Random Numbers given in Appendix A, we obtain the following ten numbers which are less than or equal to 949,

343, 613, 580, 629, 472
769, 349, 894, 112, 433.

These numbers are divided by 5 and the following quotients are obtained.

68, 122, 116, 125, 94
153, 69, 178, 22, 86.

Then, the units with serial numbers

69, 123, 117, 126, 95
154, 70, 179, 23, 87.

are selected in the sample.

This procedure is easier than Procedure 2, since in this procedure the numbers are divided by k, while in Procedure 2, the numbers are divided by N, which is a large number in comparison to k.

Procedure 4 (Independent Choice of Digits). Another method suggested by Mathai (1954) is as follows. Let N, the population size be a r-digit number. Partition N into $(N_1\ N_2)$, where N_1 is the first digit of N and N_2 consists of the remaining $(r - 1)$ digits. Now select two random numbers x and y, x from 0 to N_1, and y from $\overset{\longleftarrow r \longrightarrow}{0\ 0\ \ldots\ 0}$ to $(N_2 + 1)$. Then the unit with serial number xy is selected in the sample, if xy is neither $\overset{\longleftarrow r \longrightarrow}{0\ 0\ \ldots\ 0}$ nor greater than N. If xy is $\overset{\longleftarrow r \longrightarrow}{0\ 0\ \ldots\ 0}$ or great than N, reject it and repeat the process.

As an illustration, let $N = 190$ and $n = 10$. Then $N = 190 = N_1N_2$; hence $N_1 = 1$ and $N_2 = 90$. We select a random number x from 0 to 1 and another random number y from 00 to 91. From the page of Table of Random Numbers given in the Appendix, we take the first column of one-digit numbers for selecting x. In order to avoid more rejections, we shall consider $0, 2, 4, 6, 8$ as corresponding to number 0 and $1, 3, 5, 7, 9$ as corresponding to 1. For selecting y, we consider the first column of 2-digit numbers. We obtain the following numbers

x	y	xy
3=1	34	134
6=0	61	061
9=1	58	158
6=0	62	062
4=0	47	047
7=1	76	176
7=1	34	134*
8=0	89	089
1=1	11	111
4=0	43	043
7=1	76	176*
0=0	05	005

Here * indicates that the particular number is repeated. We wish to draw a SRSWOR; hence numbers with * are rejected. From the above table, it follows that the selected sample consists of units with serial numbers

134, 61, 158, 62, 47
176, 89, 111, 43, 5.

1.5 Determination of Sample Size

Before the actual sample survey is taken, it is necessary to determine the size of the sample. Various approaches for determination of sample size are given below.

1. Knowledge of Coefficient of Variation.

Let C be the coefficient of variation of the study variable y defined by

$$C = \frac{\sigma}{\bar{Y}},$$

where $\sigma = \sqrt{\frac{\sum_{i=1}^{N}(y_i - \bar{Y})^2}{N}} = \sqrt{\frac{(N-1)S_y^2}{N}}$. Since sampling without replacement is more efficient than samping with replacement, we consider the problem of determination of sample size for simple random sampling without replacement. We assume that C is known. We shall determine the sample size such that the relative standard error of the estimator of $\bar{Y}$ has a pre-specified value e.

For SRSWOR, we know that relative standard error of $\bar{Y}$ is

$$\frac{1}{\bar{Y}}\sqrt{\frac{N-n}{nN}S_y^2} = \sqrt{\frac{(N-n)C^2}{n(N-1)}}.$$

Equating this to e and solving the resultant equation for n, we obtain n as

$$n = \frac{Nn'}{N+n'-1}, \tag{1.5.1}$$

where $n' = \frac{C^2}{e^2}$.

2. Knowledge of Population Variance.

We assume that σ^2 is known. We also assume that n and N are fairly large. Hence $\bar{y}$ is nearly normally distributed with mean $\bar{Y}$ and standard deviation $\sigma/\sqrt{n}$. Suppose it is desired to determine the sample size so that the permissible error in the estimator of $\bar{Y}$ is d with probability $(1-\alpha)$. Thus, the value of n is required such that

$$P[|\bar{y}-\bar{Y}| \leq d] = 1-\alpha. \tag{1.5.2}$$

We can write (1.5.2) as

$$P\left[\frac{\bar{y}-\bar{Y}}{\frac{\sigma}{\sqrt{n}}} \leq \frac{d}{\sigma/\sqrt{n}}\right] = 1-\alpha. \tag{1.5.3}$$

Let $z_{\alpha/2}$ be the value of standard normal variate such that

$$P[|Z| \leq z_{\alpha/2}] = 1-\alpha. \tag{1.5.4}$$

Noting that $\frac{(\bar{y}-\bar{Y})}{\sigma/\sqrt{n}}$ is a standard normal variate, from (1.5.3) and (1.5.4) we obtain

$$n = \left[\frac{z_{\alpha/2}\sigma}{d}\right]^2. \tag{1.5.5}$$

3. Cost Aspect.

Sometimes the sample size is to be determined on the basis of the cost. Let the budget sanctioned for the survey be C'. We assume the cost function to be as

$$C' = C_0 + nC_1, \tag{1.5.6}$$

where C_0 is the overhead cost and C_1 is the cost of enumeration and processing the information per one unit. From (1.5.6), one obtains n as

$$n = \frac{C'-C_0}{C_1}. \tag{1.5.7}$$

We shall now determine the sample size so that the total cost of the survey and the loss in terms of money involved in taking a decision based on a sample of size n is minimum. For this purpose we suppose that the loss is proportional to the relative standard error of the estimator of $\bar{Y}$. Let l be the loss per 1% of the relative standard error of the estimator. Then the total survey cost and the loss involved in case of a SRSWOR is given by

$$L(n) = C_0 + nC_1 + l\sqrt{\frac{N-n}{N-1}}\frac{C}{\sqrt{n}}, \tag{1.5.8}$$

where C = coefficient of variation of the study variable y. Equating the derivative of (1.5.8) w.r.t. n to zero and after simplification we obtain

$$n^3(N-n) = \frac{(lCN)^2}{4(N-1)C_1^2}. \tag{1.5.9}$$

If N is large so that $f = n/N$ is negligible, then from (1.5.9) we obtain

$$n = \left[\frac{lC}{2C_1}\right]^{\frac{2}{3}}. \tag{1.5.10}$$

1.6 Estimation of a Proportion

Sometimes one is interested in estimating the proprtion P of units of a population belonging to a specified class. For example, we may be interested in estimating the proportion of graduates or unemployed persons in a state.

Suppose the units in the population are divided into two classes $\mathcal{C}$ and $\bar{\mathcal{C}}$, $\mathcal{C}$ consisting of units having a certain attribute. Let N_1 and N_2 be respectively the number of units in the classes $\mathcal{C}$ and $\bar{\mathcal{C}}$. Then, P, the proportion of units belong to the class $\mathcal{C}$ is

$$P = \frac{N_1}{N}, \quad N = N_1 + N_2.$$

Let $Q = 1 - P = \frac{N_2}{N}$. Our purpose is to estimate P. The results obtained in Sections 1.2 and 1.3 can be applied in estimating P, if we define the variate y as

$$\begin{aligned} y_i &= 1, \text{ if the } i^{th} \text{ unit belongs to class } \mathcal{C}, \\ &= 0, \text{ if it belongs to } \bar{\mathcal{C}}, \\ &\quad i = 1, 2, \ldots, N. \end{aligned}$$

Hence, one can verify that

$$\bar{Y} = N_1/N = P \tag{1.6.1}$$

$$S_y^2 = NPQ/(N-1) \tag{1.6.2}$$

$$C = \text{Population Coefficient of Variation}$$

$$= \sqrt{\frac{(N-1)S_y^2}{N}} \cdot \frac{1}{\bar{Y}} = \sqrt{\frac{Q}{P}}. \tag{1.6.3}$$

Suppose that a simple random sample of size n is selected and that n_1 units in this sample belong to the class $\mathcal{C}$. Let $p = \frac{n_1}{n}$ be the sample proportion of units in the class $\mathcal{C}$. Clearly,

$$\bar{y} = \frac{n_1}{n} = p,$$

$$s_y^2 = \frac{npq}{(n-1)}.$$

We now discuss two cases: (i) SRSWOR and (ii) SRSWR.

Case (i). SRSWOR. Applying the results of Theorems 1.2.1 and 1.2.3, we obtain

(i) p is an unbiased estimator of P.
(ii) $Var(p) = \frac{N-n}{N-1} \cdot \frac{PQ}{n}$.
(iii) $Est.Var(p) = \frac{N-n}{N} \cdot \frac{pq}{n-1}$.

Further, relative standard error (rse) of p is given by

(iv) $rse(p) = \frac{\sqrt{Var(p)}}{P} = \sqrt{\frac{N-n}{N-1}} \cdot \frac{1}{\sqrt{n}} \cdot \sqrt{\frac{Q}{P}}$.

Case (ii). SRSWR. Applying the results of Theorems 1.3.1 and 1.3.3, we obtain

(i) p is an unbiased estimator of P.
(ii) $Var(p) = \frac{PQ}{n}$.
(iii) $Est.Var(p) = \frac{pq}{n-1}$.

Further, the relative standard error of p is obtained as

(iv) $rse(p) = \frac{\sqrt{Var(p)}}{P} = \frac{1}{\sqrt{n}} \cdot \sqrt{\frac{Q}{P}}$.

The techniques discussed in Section 1.5 can be applied for determination of sample size in estimation of a proportion. The derivations are left to the reader.

EXERCISES

1.1. In selecting 2 units with SRSWOR from a population of $N = 5$ units with values $1, 2, 3, 4, 5$, verify that

(i) the sample mean is unbiased estimator of the population mean by enumerating all possible samples, and

(ii) the sampling variance is given by $Var(\bar{y}) = \left(\frac{1}{n} - \frac{1}{N}\right) S_y^2$ by calculating the sampling variance by enumerating all possible samples.

1.2. From a population of N units, a sample of n units is drawn in the following manner. From the population, a specified units are always selected and the remaining $(n-a)$ units are selected from the remaining $(N-a)$ units with SRSWOR. Two estimators of the population total are suggested as follows:

(i) $\hat{Y}_1 = \sum_{i=1}^{a} y_i + \frac{N-a}{n-a} \sum_{i=1}^{n-a} y_i,$

(ii) $\hat{Y}_2 = \frac{N}{n}\left(\sum_{i=1}^{a} y_i + \sum_{i=1}^{n-a} y_i\right),$

where the first a units are the specified ones. Compare the biases and variances of $\hat{Y}_1$ and $\hat{Y}_2$.

1.3. n a finite bivariate population of N units, the means and mean sum of squares of the two characteristics x and y are given by $\bar{X}$, $\bar{Y}$, S_x^2 and S_y^2 respectively, and ρ is the coefficient of correlation between x and y. A SRSWOR of size n is selected. Derive

(i) $\rho(\bar{x}, \bar{y})$, the coefficient of correlation between $\bar{x}$ and $\bar{y}$, and

(ii) $\rho(\bar{x} + \bar{y}, \bar{x} - \bar{y})$, the coefficient of correlation between $\bar{x} + \bar{y}$ and $\bar{x} - \bar{y}$.

1.4. Suppose the value of the population coefficient of variance $C = \frac{\sigma}{\bar{Y}}$ is known at the estimation stage. A simple random sample of n units is drawn with SRSWR. Can you improve upon the estimator $\bar{y}$, the sample mean? If so, suggest the improved estimator and obtain its efficiency relative to $\bar{y}$. (Hint: consider $\lambda\bar{y}$, where λ is a constant to be determined.)

1.5. A sample of n units is drawn with SRSWOR from the population, and a sub-sample of n' units is selected with SRSWOR from the first sample and added to the original sample. Derive the expected value and the approximate variance of $\bar{y}_{n+n'}$, the sample mean based on the $n + n'$ units. Determine the value of $\frac{n'}{n}$ for which the efficiency of $\bar{y}_{n+n'}$ compared to that of $\bar{y}_n$ is minimum.

1.6. A sample of 3 units is drawn from a population of N units with SRSWR. Derive the probabilities of the sample containing $d = 1, 2, 3$ distinct units. Prove that the arithmetic mean of the values of d distinct units in the sample, $\bar{y}_d$ is an unbiased estimator of the population mean $\bar{Y}$. Further obtain $Var(\bar{y}_d)$ and obtain an approximate unbiased estimator of the variance of $\bar{y}_d$.

1.7. Suppose a simple random sample of size 2 is drawn from a finite population (y_1, y_2, y_3). Corresponding to the three possible samples, $s_1 = (y_1, y_2)$, $s_2 = (y_2, y_3)$, and $s_3 = (y_1, y_3)$, a linear estimator $l(s)$ for estimating the population mean is defined as follows:

$$l(s_1) = \frac{2}{3}y_1 + \frac{1}{2}y_2$$

$$l(s_2) = \frac{1}{2}y_2 + \frac{1}{2}y_3$$

$$l(s_3) = \frac{1}{3}y_1 + \frac{1}{2}y_3.$$

(a) Show that the estimator $l(s)$ is an unbiased estimator of the population mean and obtain its variance.

(b) Derive the condition on the values of y_1, y_2, y_3 for which the $Var(l(s))$ is less than that of the sample mean $\bar{y}$. Hence, in particular, show that, if $y_1 = 1$, $y_2 = 2$, $y_3 = 3$, $Var(l(s)) < Var(\bar{y})$.

1.8. A simple random sample of size n is drawn without replacement from a population of size N. A linear estimator of the population mean $\bar{Y}_N$ is defined as

$$\widehat{\bar{Y}_N} = \sum_{i=1}^{n} \lambda_i y_i',$$

where λ's are certain constants and y_i' denotes the value of the unit included in the sample at the ith draw.

(a) Prove that $\widehat{\bar{Y}_N}$ is an unbiased estimator of $\bar{Y}_N$ if and only if $\sum_{i=1}^{n} \lambda_i = 1$ and that when $\sum_{i=1}^{n} \lambda_i = 1$, its variance is given by

$$Var(\widehat{\bar{Y}_N}) = \frac{S_y^2}{N}\left[N\sum_{i=1}^{n} \lambda_i^2 - 1\right],$$

where S_y^2 is the mean sum of squares of the population.

(b) Prove that, when $\sum_{i=1}^{n} \lambda_i = 1$, the variance of $\widehat{\bar{Y}_N}$ is minimized if $\lambda_i = \frac{1}{n}$, $i = 1, 2, \ldots, n$.

1.9. A simple random sample of size m (fixed in advance) is drawn with replacement from a population of N units. Let d denote the number of distinct units in the sample, the ith unit occuring r_i times with $\sum_{i=1}^{d} r_i = m$. The following two estimators of the population mean are defined:

$$\bar{y}_m = \frac{\sum_{i=1}^{m} r_i y_i}{m}, \qquad \bar{y}_d = \frac{\sum_{i=1}^{d} y_i}{d},$$

where y_i is the value of the ith unit in the sample.

(i) Show that $\bar{y}_m$ and $\bar{y}_d$ are unbiased and that $\bar{y}_d$ is more efficient that $\bar{y}_m$ if

$$E\left(\frac{1}{d}\right) < \frac{1}{m}\left(1 + \frac{m-1}{n}\right).$$

(ii) Show that the probability distribution of d is given by

$$P(d) = \frac{1}{N^m}\binom{N}{d}\sum_{t=0}^{d}(-1)^{d-t} \cdot \binom{d}{t} t^m.$$

(iii) Hence, or otherwise find $E(d)$ and $E\left(\frac{1}{d}\right)$.

1.10. For the estimator $\bar{y}_d$ defined in Exercise 1.9, obtain $Var(\bar{y}_d)$, and show that its unbiased estimator is given by

$$Est.Var(\bar{y}_d) = \left(\frac{1}{n} - \frac{1}{N}\right) s^2,$$

where s^2 is the sample mean sum of squares based on d distinct units.

1.11. A simple random sample of n distinct units (n fixed in advance) is selected with replacement from a population of N units. Denote the total sample size including repetitions by the random variable v. Let the ith distinct unit occur r_i times in the sample so that $\sum_{i=1}^{n} r_i = v$. The following two estimators are defined:

$$\bar{y}_v = \frac{1}{v}\sum_{i=1}^{n} r_i y_i, \qquad \bar{y}_n = \frac{1}{n}\sum_{i=1}^{n} y_i.$$

(a) Show that $\bar{y}_v$ and $\bar{y}_n$ are unbiased and that $\bar{y}_n$ is more efficient than $\bar{y}_v$ if

$$E\left(\frac{1}{v}\right) > \frac{1}{n} \cdot \frac{N-n}{N-1}.$$

(b) Show that the probability distribution of v is given by

$$P(v) = \binom{N-1}{n-1} N^{1-v} \sum_{t=0}^{n-1} (-1)^{n-1-t} \binom{n-1}{t} t^{v-1}.$$

(c) Hence, or otherwise find $E(v)$ and $E\left(\frac{1}{v}\right)$.

1.12. Suppose there are two lists, one having M units and the other having N units, and it is required to estimate the total number of units D common to both the lists. For this purpose, simple random samples of m and n units, selected without replacement from the two lists are compared and it is found that d units are common between the two samples.

(i) Give an unbiased estimator of D and obtain its variance.

(ii) Show that the probability distribution of d is given by

$$P(d) = \frac{\binom{D}{d}}{\binom{M}{m}\binom{N}{n}} \sum_{k=d}^{D} \binom{D-d}{k-d} \binom{M-D}{m-k} / \binom{N-k}{n-d}$$

(iii) If M, N, m and $n \to \infty$ such that D, $\frac{m}{M}$, $\frac{n}{N}$ remain constant, show that

$$P(d) \to \binom{D}{d} f^d (1-f)^{D-d}$$

where $f = mn/MN$.

(iv) If M, N, m, n and $D \to \infty$ such that $\frac{mnD}{MN}$ remains constant, λ say, then show that

$$P(d) \to \frac{\lambda^d e^{-\lambda}}{d!}.$$

1.13. A SRSWOR of size $n = n_1 + n_2$ is drawn from a population of N units and $\bar{y}_n$ is the mean of it. A SRSWOR of size n_1 is drawn from it with mean $\bar{y}_{n_1}$. Show that

(i) $Var(\bar{y}_{n_1} - \bar{y}_{n_2}) = s^2 \left[\frac{1}{n_1} + \frac{1}{n_2}\right]$, where $\bar{y}_{n_2}$ is the mean of the remaining n_2 units in the sample,

(ii) $Var(\bar{y}_{n_1} - \bar{y}_n) = s^2 \left[\frac{1}{n_1} - \frac{1}{n}\right]$,

(iii) $Cov(\bar{y}_n, \bar{y}_{n_1} - \bar{y}_n) = 0$.

1.14. A simple random sample of size 3 is drawn from a population of N units with replacement. Show that the probabilities that the sample contains 1, 2, and 3 different units are

$$P_1 = \frac{1}{N^2}, \qquad P_2 = \frac{3(N-1)}{N^2}, \qquad P_3 = \frac{(N-1)(N-2)}{N^2}.$$

An estimator of the population mean $\bar{Y}$ is taken as $\bar{y}^*$, the unweighted mean over the different units in the sample. Show that the average variance of $\bar{y}^*$ is given by

$$Var(\bar{y}^*) = \frac{(2N-1)(N-1)S_y^2}{6N^2} \cong \left(1 - \frac{f}{2}\right) S_y^2/3.$$

This is shown by obtaining

$$Var(\bar{y}^*) = S_y^2 \left[\frac{N-1}{N} P_1 + \frac{N-2}{2N} P_2 + \frac{N-3}{3N} P_3\right].$$

Hence, show that $Var(\bar{y}^*) < Var(\bar{y})$, where $\bar{y}$ is the ordinary mean of 3 observations of the sample.

1.15. Suppose it is realized that y_1 would be unusually low and y_N unusually high. Consider the following estimator of $\bar{Y}$, with SRSWOR.

$$\begin{aligned} \hat{Y}_s &= \bar{y} + c, && \text{if the sample contains } y_1 \text{ but not } y_N \\ &= \bar{y} - c, && \text{if the sample contains } y_N \text{ but not } y_1 \\ &= \bar{y}, && \text{for all other samples} \end{aligned} \tag{1.6.1}$$

where c is some constant. Prove that $\hat{Y}_s$ is an unbiased estimator of $\bar{Y}$ and that

$$Var(\hat{Y}_s) = (1-f)\left[\frac{S_y^2}{n} - \frac{2c}{N-1}(y_N - y_1 - nc)\right]$$

and hence prove that

$$Var(\hat{Y}_s) < Var(\bar{y}) \text{ if } 0 < c < \frac{(y_N - y_1)}{n}.$$

1.16. In Exercise 1.15, let $N = 8$ and the y values be $y_1 = 1, 4, 5, 6, 6, 8, 13 = y_N$. For $n = 4$, show that $Var(\bar{y}) = 1.5$, while $Var(\hat{Y}_s) = 0.214$, when $c = 1.5$ (its best value) and $Var(\hat{Y}_s) = 0.357$ when $c = 1$ or 2.

1.17. Suppose that y_1 is usually low and y_N is usually high. Consider a sampling plan as follows. Always include both y_1 and y_N in the sample and draw a SRSWOR of size $(n-2)$ from $y_2, y_3, \ldots, y_{N-1}$. Let the mean of this SRSWOR be $\bar{y}_{n-2}$. Show that

$$\hat{\hat{Y}} = \frac{y_1 + (N-2)\bar{y}_{n-2} + y_N}{N}$$

is an unbiased estimator of $\bar{Y}$. Obtain the variance of $\hat{\hat{Y}}$.

1.18. Suppose a population has N units and the value of one unit is known to be y_0. A SRSWOR of n units is selected from the remaining $(N-1)$ units. Show that the estimator $y_0 + (N-1)\bar{y}_n$ is unbiased for the population total Y. Derive the variance of this estimator and show that its variance is less than that of $N\bar{y}_n$ based on a SRSWOR of size n from the entire population.

Chapter 2

SAMPLING WITH VARYING PROBABILITIES OF SELECTION

2.1 Introduction

In Chapter 1, we discussed sampling in which every unit of the population is selected with equal probability. Sometimes more efficient estimators are obtained by assigning unequal probabilities of selection of the units of the population. This type of sampling in which unequal probabilities of selection are assigned to the units of the population is called *sampling with varying probabilities of selection*. The most commonly used sampling with varying probabilities of selection is the sampling in which units are selected with probability proportional to the size of the unit. The size being the value of an auxiliary variable x related to the study variable y. This sampling scheme is known as pps sampling. For example, in an industrial survey, the number of workers may be taken as the size of an industrial establishment. In estimating crop yeild, the cultiviated area for a previous period may be taken as the size.

2.2 Sampling with Varying Probabilities of Selection and with Replacement

Let there be N units in the population and P_i be the probability of selection assigned to the ith unit of the population, $i = 1, 2, \ldots, N$. Let y_i be the value of the study variable y associated with the ith unit of the population, $i = 1, 2, \ldots, N$. As usual, $\bar{Y} = \sum_{i=1}^{N} y_i/N$ and $Y = \sum_{i=1}^{N} y_i$ represent respectively the population mean and the population total. We consider a sample of size n with replacement.

Theorem 2.2.1. *In sampling with varying probabilities of selection and with replacement,*

(i) $y* = \frac{1}{n}\sum_{i=i}^{n}(y_i/NP_i)$ is an unbiased estimator of $\bar{Y}$,

(ii)

$$Var(y*) = \frac{1}{nN^2}\left[\sum_{i=1}^{N} y_i^2/P_i - N^2\bar{Y}^2\right] = \frac{1}{nN^2}\sum_{i=1}^{N} P_i\left(\frac{y_i}{P_i} - Y\right)^2$$

(iii) an unbiased estimator of variance of $y*$ is given by

$$Est.Var(y*) = \frac{1}{n(n-1)}\sum_{i=i}^{n}\left(\frac{y_i}{NP_i} - y*\right)^2.$$

Proof. We define a_i as the number of times the ith unit occurs in the sample, $i = 1, 2, \ldots, N$. Clearly, a_i assumes values $0, 1, 2, \ldots, n$. The joint probability distribution of $a_1, a_2, \ldots, a_N$ is multinomial with

$$E(a_i) = nP_i, \qquad Var(a_i) = nP_i(1 - P_i)$$
$$Cov(a_i, a_j) = -nP_iP_j.$$

Now

$$y* = \frac{1}{nN}\sum_{i=1}^{n} y_i/P_i = \frac{1}{nN}\sum_{i=1}^{N} a_i y_i/P_i.$$

Hence,

$$E(y*) = \frac{1}{nN}\sum_{i=1}^{N} E(a_i)y_i/P_i = \frac{1}{N}\sum_{i=1}^{N} y_i = \bar{Y}.$$

This proves part (i) of the theorem. Further

$$Var(y*) = \frac{1}{n^2N^2}Var\left(\sum_{i=1}^{N} a_i y_i/P_i\right)$$
$$= \frac{1}{n^2N^2}\left[\sum_{i=1}^{N} Var(a_i)y_i^2/P_i^2 + \sum_{i\neq j}^{N} Cov(a_i, a_j)y_i y_j/P_iP_j\right]. \tag{2.2.1}$$

Substituting the values of $Var(a_i)$ and $Cov(a_i, a_j)$ in (2.2.1) and simplifying, we obtain

$$Var(y*) = \frac{1}{nN^2}\left[\sum_{i=1}^{N} y_i^2/P_i - N^2\bar{Y}^2\right]. \tag{2.2.2}$$

This proves part (ii) of the theorem since (2.2.2) can be expressed as

$$Var(y*) = \frac{1}{nN^2} \sum_{i=1}^{N} P_i \left(\frac{y_i}{P_i} - Y \right)^2 .$$

Lastly, we have

$$\frac{1}{n(n-1)} \sum_{i=1}^{n} \left(\frac{y_i}{NP_i} - y* \right)^2 = \frac{1}{n(n-1)} \left[\sum_{i=1}^{n} \frac{y_i^2}{N^2 P_i^2} - ny*^2 \right]$$

$$= \frac{1}{n(n-1)} \left[\sum_{i=1}^{N} \frac{a_i y_i^2}{N^2 P_i^2} - ny*^2 \right] .$$

Hence, taking expectation, we get

$$E \left[\frac{1}{n(n-1)} \sum_{i=1}^{N} \left(\frac{y_i}{NP_i} - y* \right)^2 \right] = \frac{1}{n(n-1)} \left[\sum_{i=1}^{N} \frac{ny_i^2}{N^2 P_i} - nE(y*^2) \right]$$

$$= \frac{1}{(n-1)} \left[\sum_{i=1}^{N} y_i^2 / N^2 P_i - Var(y*) - \bar{Y}^2 \right]$$

$$= \frac{1}{(n-1)} [nVar(y*) - Var(y*)]$$

$$= Var(y*),$$

which proves part (iii) of the theorem.

Remark. If we take $P_i = 1/N$, then from Theorem 2.2.1, we obtain the results for SRSWR established in Theorems 1.3.1 and 1.3.3.

2.3 Comparison with SRSWR

In this section, we shall compare sampling with varying probabilities of selection and with replacement with SRSWR.

In SRSWR, the variance of $\bar{y}$, the unbiased estimtor of $\bar{Y}$ is given by

$$Var(\bar{y}) = \frac{N-1}{nN} S_y^2 = \frac{1}{nN} \left[\sum_{i=1}^{N} y_i^2 - N\bar{Y}^2 \right] .$$

In sampling with varying probabilities of selection and with replacement, the variance of $y*$, the unbiased estimator of $\bar{Y}$ is given by

$$Var(y*) = \frac{1}{nN^2} \left[\sum_{i=1}^{N} y_i^2 / P_i - N^2 \bar{Y}^2 \right] . \tag{2.3.2}$$

From (2.3.1) and (2.3.2), we get

$$Var(\bar{y}) - Var(y*) = \frac{1}{nN^2}\sum_{i=1}^{N} y_i^2\left(N - \frac{1}{P_i}\right). \tag{2.3.3}$$

Hence, the gain in efficiency due to sampling with varying probabilities of selection and with replacement over SRSWR is given by

$$G = \frac{\frac{1}{nN^2}\sum_{i=1}^{N} y_i^2\left(N - \frac{1}{P_i}\right)}{\frac{1}{nN^2}\left[\sum_{i=1}^{N} y_i^2/P_i - N^2\bar{Y}^2\right]}. \tag{2.3.4}$$

We can estimate (2.3.3) by $\frac{1}{n^2N^2}\sum_{i=1}^{N}\frac{y_i^2}{P_i}\left(N - \frac{1}{P_i}\right)$ from the sample drawn with varying probabilities of selection and with replacement, since

$$E\left[\frac{1}{n^2N^2}\sum_{i=1}^{n}\frac{y_i^2}{P_i}\left(N - \frac{1}{P_i}\right)\right] = E\left[\frac{1}{n^2N^2}\sum_{i=1}^{N} a_i\frac{y_i^2}{P_i}\left(N - \frac{1}{P_i}\right)\right]$$
$$= \frac{1}{nN^2}\sum_{i=1}^{N} y_i^2\left(N - \frac{1}{P_i}\right)$$

where a_i's are as defined in the proof of Theorem 2.2.1.

Also $Var(y*)$ is estimated by $\frac{1}{n(n-1)}\sum_{i=1}^{n}\left(\frac{y_i}{NP_i} - y*\right)^2$. Hence, G is estimated by

$$\hat{G} = \frac{\frac{1}{n^2N^2}\sum_{i=1}^{n}\frac{y_i^2}{P_i}\left(N - \frac{1}{P_i}\right)}{\frac{1}{n(n-1)}\sum_{i=1}^{n}\left(\frac{y_i}{NP_i} - y*\right)^2}. \tag{2.3.5}$$

2.4 Sampling with Varying Probabilities of Selection and without Replacement

Suppose that there are N units in the population. Let P_{i_r} be the probability that the ith unit is selected in the sample at the rht draw. It is assumed that we make n draws without replacement. Hence, in P_{i_r}, $i = 1, 2, \ldots, N$; $r = 1, 2, \ldots, n$. For convenience, we shall write P_i for P_{i_1}.

Let us define a_i for $i = 1, 2, \ldots, N$ as follows.

$$a_i = 1, \quad \text{if the } i^{th} \text{ unit occurs in the sample.}$$
$$= 0, \quad \text{if the } i^{th} \text{ unit does not occur in the sample.}$$

Then, clearly

$$E(a_i) = \text{Probability that the } i^{th} \text{ unit is included in the sample}$$
$$= \sum_{r=1}^{n} P_{i_r} = \pi_i$$

$$E(a_i a_j) = \text{Probability that the } i^{th} \text{ and } j^{th} \text{ units are included in the sample}$$
$$= \pi_{ij}, \quad i \neq j = 1, 2, \ldots, N.$$

π_i and π_{ij} are known as the inclusion probabilities for the ith unit and a pair of units i and j. In the following theorem we give the estimator of $\bar{Y}$ proposed by Horvitz and Thompson (1952).

Theorem 2.4.1. *In sampling with varying probabilities of selection and without replacement,*

(i) $y^*_{HT} = \frac{1}{N}\sum_{i=1}^{n} y_i/E(a_i)$ *is an unbiased estimator of the population mean* $\bar{Y}$.

(ii)

$$Var(y^*_{HT}) = \frac{1}{N^2}\left[\sum_{i=1}^{N} \frac{1 - E(a_i)}{E(a_i)} y_i^2 + \sum_{i \neq j}^{N} \frac{E(a_i a_j) - E(a_i)E(a_j)}{E(a_i)E(a_j)} y_i y_j\right].$$

(iii) *An unbiased estimator of variance of* y^*_{HT} is given by

$$Est.Var(y^*_{HT}) = \frac{1}{N^2}\left[\sum_{i=1}^{n} \frac{1 - E(a_i)}{E^2(a_i)} y_i^2 + \sum_{i \neq j}^{n} \frac{E(a_i a_j) - E(a_i)E(a_j)}{E(a_i a_j)E(a_i)E(a_j)} y_i y_j\right].$$

Proof. We can write y^*_{HT} as

$$y^*_{HT} = \frac{1}{N}\sum_{i=1}^{N} a_i y_i / E(a_i).$$

Hence, we get

$$E(y^*_{HT}) = \frac{1}{N}\sum_{i=1}^{N} y_i = \bar{Y},$$

which prove part (i) of the theorem. Further,

$$Var(y^*_{HT}) = \frac{1}{N^2}\left[\sum_{i=1}^{N} Var(a_i) y_i^2 / E^2(a_i) + \sum_{i \neq j}^{N} Cov(a_i, a_j) y_i y_j / E(a_i)E(a_j)\right]. \tag{2.4.1}$$

Now, noting that

$$Var(a_i) = E(a_i)(1 - E(a_i))$$
$$Cov(a_i, a_j) = E(a_i a_j) - E(a_i)E(a_j),$$

we get from (2.4.1),

$$Var(y^*_{HT}) = \frac{1}{N^2}\left[\sum_{i=1}^{N} \frac{1 - E(a_i)}{E(a_i)} y_i^2 + \sum_{i \neq j}^{N} \frac{E(a_i a_j) - E(a_i)E(a_j)}{E(a_i)E(a_j)} y_i y_j\right]. \tag{2.4.2}$$

This proves part (ii) of the theorem. Lastly,

$$E\left[\frac{1}{N^2}\left\{\sum_{i=1}^{n} \frac{1 - E(a_i)}{E^2(a_i)} y_i^2 + \sum_{i \neq j}^{n} \frac{E(a_i a_j) - E(a_i)E(a_j)}{E(a_i a_j)E(a_i)E(a_j)} y_i y_j\right\}\right]$$
$$= E\left[\frac{1}{N^2}\left\{\sum_{i=1}^{N} \frac{a_i(1 - E(a_i))}{E^2(a_i)} y_i^2 + \sum_{i \neq j}^{N} \frac{a_i a_j (E(a_i a_j) - E(a_i)E(a_j))}{E(a_i a_j)E(a_i)E(a_j)} y_i y_j\right\}\right]$$
$$= \frac{1}{N^2}\left[\sum_{i=1}^{N} \frac{1 - E(a_i)}{E(a_i)} y_i^2 + \sum_{i \neq j}^{N} \frac{E(a_i a_j) - E(a_i)E(a_j)}{E(a_i)E(a_j)} y_i y_j\right]$$
$$= Var(y^*_{HT}),$$

which proves part (iii) of the theorem.

Remark 1. We can write the result of Theorem 2.4.1 in terms of inclusion probabilities as follows:

(i) $y^*_{HT} = \frac{1}{N}\sum_{i}^{n} y_i/\pi_i$ is an unbiased estimator of $\bar{Y}$.

(ii)

$$Var(y^*_{HT}) = \frac{1}{N^2}\left[\sum_{i=1}^{N} \frac{1 - \pi_i}{\pi_i} y_i^2 + \sum_{i \neq j}^{N} \frac{\pi_{ij} - \pi_i \pi_j}{\pi_i \pi_j} y_i y_j\right].$$

(iii) An unbiased estimator of variance of y^*_{HT} is given by (due to Horvitz and Thompson)

$$Est.Var(y^*_{HT})_{HT} = \frac{1}{N^2}\left[\sum_{i=1}^{n} \frac{1 - \pi_i}{\pi_i^2} y_i^2 + \sum_{i \neq j}^{n} \frac{\pi_{ij} - \pi_i \pi_j}{\pi_{ij}\pi_i \pi_j} y_i y_j\right].$$

Remark 2. For SRSWOR, $\pi_i = n/N$, $\pi_{ij} = \frac{n(n-1)}{N(N-1)}$. Hence as the results of SRSWOR follow from Theorem 2.4.1 as a particular case. The derivatives are left as an exercise to the reader. Another elegant form of the variance of y^*_{HT} is due to Yates and Grundy (1953) which is given in Theorem 2.4.2.

Theorem 2.4.2. *The Yates and Grundy form of $Var(y^*_{HT})$ is given by*

$$Var(y^*_{HT})_{YG} = \frac{1}{2N^2}\sum_{i\neq j}^{N}(\pi_i\pi_j - \pi_{ij})\left[\frac{y_i}{\pi_i} - \frac{y_i}{\pi_j}\right]^2.$$

Proof. We have

$$\frac{1}{2N^2}\sum_{i\neq j}^{N}(\pi_i\pi_j - \pi_{ij})\left(\frac{y_i}{\pi_i} - \frac{y_i}{\pi_j}\right)^2$$

$$= \frac{1}{2N^2}\sum_{i\neq j}^{N}(\pi_i\pi_j - \pi_{ij})\frac{y_i^2}{\pi_i^2} - \frac{1}{N^2}\sum_{i\neq j}^{N}\frac{\pi_i\pi_j - \pi_{ij}}{\pi_i\pi_j}y_iy_j$$

$$+ \frac{1}{2N^2}\sum_{i\neq j}^{N}(\pi_i\pi_j - \pi_{ij})\frac{y_i^2}{\pi_j^2}. \tag{2.4.3}$$

Consider the coefficient of y_i^2/π_i^2 in the first term on the r.h.s. of (2.4.3), which is

$$\frac{1}{2N^2}\pi_i\left(\sum_{j=1}^{N}\pi_j - \pi_i\right) - \frac{1}{2N^2}\sum_{j\neq i}^{N}\pi_{ij}.$$

Now, we know that $\sum_{j=1}^{N} a_j = n$, hence,

$$\sum_{j=1}^{N} E(a_j) = n, \quad \text{i.e.} \quad \sum_{j=1}^{N}\pi_j = n.$$

Also, $\sum_{j=1}^{N} a_i a_j = n a_i$, hence,

$$\sum_{j=1}^{N} E(a_i a_j) = nE(a_i)$$

i.e.

$$\sum_{j=1}^{N} = \pi_{ij} = n\pi_i$$

i.e.

$$\sum_{j\neq i}^{N}\pi_{ij}=n\pi_i-\pi_i \quad (\text{therefore } \pi_{ii}=\pi_i)=(n-1)\pi_i.$$

Thus, the coefficient of y_i^2/π_i^2 in the first term of r.h.s. of (2.4.3) is

$$\frac{1}{2N^2}\pi_i(n-\pi_i)-\frac{1}{2N^2}(n-1)\pi=\frac{1}{2N^2}\pi_i(1-\pi_i).$$

Therefore, the first term on the r.h.s. of (2.4.3) becomes

$$\frac{1}{2N^2}\sum_{i=1}^{N}\frac{1-\pi_i}{\pi_i}y_i^2.$$

Similarly, the third term on the r.h.s of (2.4.3) becomes

$$\frac{1}{2N^2}\sum_{i=1}^{N}\frac{1-\pi_i}{\pi_i}y_i^2,$$

and the r.h.s. of (2.4.3) becomes

$$\frac{1}{N^2}\left[\sum_{i=1}^{N}\frac{1-\pi_i}{\pi_i}y_i^2+\sum_{i\neq j}^{N}\frac{\pi_{ij}-\pi_i\pi_j}{\pi_i\pi_j}y_iy_j\right]=Var(y_{HT}^*).$$

Thus, the theorem is proved.

Theorem 2.4.3. *An unbiased estimator of $Var(y_{HT}^*)_{YG}$ is given by*

$$Est.Var(y_{HT}^*)_{YG}=\frac{1}{2N^2}\sum_{i\neq j}^{n}\frac{\pi_i\pi_j-\pi_{ij}}{\pi_{ij}}\left(\frac{y_i}{\pi_i}-\frac{y_i}{\pi_j}\right)^2.$$

Proof.

$$E\left[\frac{1}{2N^2}\sum_{i\neq j}^{n}\frac{\pi_i\pi_j-\pi_{ij}}{\pi_{ij}}\left(\frac{y_i}{\pi_i}-\frac{y_i}{\pi_j}\right)^2\right]$$

$$=E\left[\frac{1}{2N^2}\sum_{i\neq j}^{N}\frac{a_ia_j(\pi_i\pi_j-\pi_{ij})}{\pi_{ij}}\left(\frac{y_i}{\pi_i}-\frac{y_i}{\pi_j}\right)^2\right]$$

$$=\frac{1}{2N^2}\sum_{i\neq j}^{N}\frac{\pi_i\pi_j-\pi_{ij}}{\pi_{ij}}\left(\frac{y_i}{\pi_i}-\frac{y_j}{\pi_j}\right)^2$$

$$=Var(y_{HT}^*)_{YG},$$

which proves the theorem.

Thus, we have two equivalent expressions for the unbiased estimator of $Var(y_{HT}^*)$, one due to Horvitz and Thompson and another due to Yates and Grundy. The main disadvantage of these variance estimators is that they take negative values for some samples and this leads to difficulties in interpreting the reliability of estimates.

2.5 Midzuno Scheme of Sampling

The sampling scheme proposed by Midzuno (1951) consists in selecting the first unit with unequal probabilities of selection and then drawing a sample of $(n-1)$ units from the remaining $(N-1)$ units with SRSWOR. Under this scheme of sampling, we compute π_i and π_{ij} as follows.

$$\begin{aligned}
\pi_i &= \text{Prob}(i^{th} \text{ unit is included in the sample}) \\
&= \text{Prob}(i^{th} \text{ unit is included in the sample at the first draw}) \\
&\quad + \text{Prob}(i^{th} \text{ unit is not included in the sample at the first sample}) \\
&\quad \times \text{Prob}(i^{th} \text{ unit is included in the sample at any of the subsequent} \\
&\quad\quad \text{draws}) \\
&= P_i + (1-P_i)\frac{(n-1)}{(N-1)} \\
&= \frac{n-1}{N-1} + \frac{N-n}{N-1}P_i.
\end{aligned}$$

$$\begin{aligned}
\pi_{ij} &= \text{Prob that } i^{th} \text{ and } j^{th} \text{ units are included in the sample} \\
&= \text{Prob. that the } i^{th} \text{ unit is selected at the first draw} \\
&\quad\quad \text{and the } j^{th} \text{ unit is selected at any of the subsequent draws} \\
&\quad + \text{Prob. that the } j^{th} \text{ unit is selected at the first draw} \\
&\quad\quad \text{and the } i^{th} \text{ unit is selected at any of the subsequent draws} \\
&\quad + \text{Prob. that neither } i^{th} \text{ unit nor } j^{th} \text{ unit is selected at the} \\
&\quad\quad \text{first draw and they are selected at subsequent draws} \\
&= P_i \cdot \frac{n-1}{N-1} + P_j \cdot \frac{n-1}{N-1} + (1-P_i-P_j)\frac{(n-1)(n-2)}{(N-1)(N-2)} \\
&= \frac{(n-1)(n-2)}{(N-1)(N-2)} + \frac{(n-1)(N-n)}{(N-1)(N-2)}(P_i+P_j).
\end{aligned}$$

Substituting the values of π_i and π_{ij} in $\pi_i\pi_j - \pi_{ij}$, we obtain after some simplification

$$\pi_i\pi_j - \pi_{ij} = \frac{(N-n)(n-1)}{(N-1)^2(N-2)}(1-P_i-P_j) + \frac{(N-n)^2}{(N-1)^2}P_iP_j.$$

Hence, under Midzuno scheme of sampling, the Yates and Grundy estimator of $Var(y^*_{HT})$ is always non-negative.

We may note that the probability $P(s)$ of getting a particular unordered sample s is the sum of probabilities of getting it with the sample unit

u_i, $(i = 1, 2, \ldots, n)$ being selected at the first draw. Hence

$$P(s) = \sum_{i}^{n} P_i \cdot \frac{1}{\binom{N-1}{n-1}},$$

since the probability of selecting the unit u_i at the first draw is P_i and that of selecting the remaining $(n-1)$ units of the sample is $\frac{1}{\binom{N-1}{n-1}}$.

2.6 pps Sampling with Replacement

One of the most widely used methods of sampling with varying probabilities of selection is the sampling in which units are selected with probabilities proportional to their size at any draw. This type of sampling is called as the pps (probability proportional to size) sampling. This size is the value of some auxiliary variable x correlated with the study variable y.

In pps sampling with replacement, P_i, the probability of selection of ith unit at first draw remain same at every draw. Let x_i, be the value of the auxiliary variable associated with the ith unit, $i = 1, 2, \ldots, N$ and $X = \sum_{i=1}^{N} x_i$ be its total. Then, in pps sampling with replacement, $P_i = x_i/X$. Then, using the results of Theorem 2.2.1, we have the following.

Theorem 2.6.1. *In pps sampling with replacement,*

(i) $y^* = \bar{X}\bar{r}$ *is an unbiased estimator of* $\bar{Y}$, *where* $n\bar{r} = \sum^{n} r_i$, *and* $r_i = y_i/x_i$.

(ii) $Var(y^*) = \frac{1}{nN}\left[\bar{X}\sum_{i=1}^{N} y_i^2/x_i - N\bar{Y}^2\right]$,

(iii) $Est.Var(y^*) = \frac{\bar{X}^2}{n(n-1)}\sum_{i=1}^{n}(r_i - \bar{r})^2$.

We note that the estimated variance is non-negative here.

2.7 pps Sampling without Replacement

Here we shall select a sample of n units from a population of N units with probability proportional to size measure x at each draw without replacing the units selected in the previous draws. The probability of selection of the ith unit at the first draw is given by $P_i = x_i/X$, $i = 1, 2, \ldots, N$. Given that the ith unit is selected at the first draw, the probability of selection of the jth unit at the second draw is $P_j/(1 - P_i)$, $j \neq i = 1, 2, \ldots, N$. Given that the ith and j^{th} units are selected in the first two draws, the probability of selection the kth unit at the third draw is $P_k/(1 - P_i - P_j)$, $k = i = j = 1, 2, \ldots, N$ and so on. The results of Section 2.4 then apply.

One of the important problems in sampling with varying probabilities of selection without replacement is the determination of probabilities P_i' to be used at different draws such that P_i becomes proportional to x_i, Narain (1951), Yates and Grundy (1953), Hanurav (1962a,b), and Fellegi (1964) have given procedures of determining probabilities P_i' to be used at different draws such that P_i becomes proportional to x_i. But these procedures are more complicated and time-consuming. Hence they are not given here.

2.8 Selection Procedures for pps Sampling

Here we shall describe two procedures for selecting a pps sample. The two procedures are (1) Cumulative Totals Method, and (2) Lahiri's Method.

(1) Cumulative Totals Method

Let the size of the ith unit be x_i, $i = 1, 2, \ldots, N$. The steps involved in using this method are as follows:

(i) Write down the cumulative totals defined by

$$\begin{aligned} T_i &= x_1 + x_2 + \ldots + x_{i-1} + x_i \\ &= T_{i-1} + x_i, \quad i = 1, 2, \ldots, N \\ T_0 &= 0 \text{ by definition} \\ T_N &= X. \end{aligned}$$

(ii) Choose a random number from 1 to T_N. Let it be R.

(iii) If $T_{i-1} < R \leq T_i$, then select the ith unit. Thus, we see that the probability of selecting the ith unit is $\frac{T_i - T_{i-1}}{T_N} = \frac{x_i}{X}$, which is proportional to its size. For selecting a pps sample of size n with replacement, the above procedure is repeated n times. For selecting a pps sample of size n without replacement, the first unit is selected by the above procedure and then it is deleted from the population and for the remainder population, new cumulative totals are calculated and again the same procedure is used to select a second unit. The procedure is continued until a sample of n units is obtained.

Example 2.1. The procedure is illustrated in Table 2.1. There are 10 factories. Number of workers in a factory is taken as the size variable x.

Table 2.1

Factory No.	No. of Workers	Cumulative Total
(1)	(2)	(3)
1	78	78
2	150	228
3	356	584
4	112	696
5	200	896
6	550	1446
7	444	1890
8	160	2050
9	666	2716
10	284	3000

Table 2.1 gives the factories and number of workers for each factory. The third column gives the cumulative totals.

Suppose we wish to draw a pps sample of 3 factories with replacement. From the page of table of random numbers given in Appendix A, we select three numbers which are less than 3000, from the left column of 4-digited numbers. These numbers are: $1122, 0592, 2630$. Now $896 < 1122 < 1446$, hence the factory No. 6 is selected. $584 < 592 < 696$, hence factory No. 4 is selected. Also, $2050 < 2630 < 2716$, hence factory No. 9 is selected. Hence, the pps sample of 3 factories with replacement consists of factories with serial numbers $4, 6, 9$.

Suppose we wish to draw a pps sample of 3 factories without replacement. For the selection of first unit, the above procedure is followed, and hence factory No. 6 is selected. Now we remove factory No. 6 from the population and calculate the cumulative totals again. These are given in Table 2.2.

From the page of table of random numbers given in Appendix A, we

Table 2.2

Factory No.	No. of Workers	Cumulative Total
1	78	78
2	150	228
3	356	584
4	112	696
5	200	896
7	444	1340
8	160	1500
9	666	2166
10	284	2450

Table 2.3

Factory No.	No. of Workers	Cumulative Total
1	78	78
2	150	228
3	356	584
4	112	696
5	200	896
8	160	1056
9	666	1722
10	284	2006

select a number which is less than or equal to 2450 from the second column of 4-digited numbers. The number is 0924. Since $896 < 924 < 1340$, factory No. 7 is selected at the second draw. Now we remove factory No. 7 from the population and calculate new cumulative totals. These are given in Table 2.3.

From the page of table of random numbers given in Appendix A, we select a number which is less than or equal to 2006 from the third column of 4-digited numbers. This number is 1298. Since $1056 < 1298 < 1722$, factory No. 9 is selected.

Thus, the pps sample of 3 factories without replacement consists of factories with serial numbers $6, 7, 9$.

The main disadvantage of the cumulative total method is that it involves writing down of the cumulative totals T_i which is tedious and time consuming when N is large.

(2) Lahiri's Method

Lahiri (1951) suggested a method for selecting a pps sample which does not require the writing down of cumulative totals. Lahiri's method consists of the following steps:

(i) Select a number from 1 to N. Let it be i, say.

(ii) Select a number from 1 to M, where M is a convenient number greater than or equal to the maximum size. Let it be R, say.

(iii) If $R \leq x_i$, then i^{th} unit is selected. If $R > x_i$, then i^{th} unit is rejected and the procedure is repeated.

For selection of a pps sample of n units with replacement, the above procedure is repeated till n units are selected.

For selection of a pps sample of n units without replacement, the units selected in earlier draws are removed from the population, and Lahiri's method is used for selecting a unit at the next draw.

Example 2.2. We use the data of Example 2.1. Here $N = 10$. The maximum size is 666. Hence we take $M = 1,000$.

Suppose we wish to select a pps sample of 3 factories with replacement. We select a pair of numbers (i, R) such that $1 \leq i \leq 10$ and $1 \leq R \leq 1,000$. for selection of random numbers, we use Procedure 2 given in Section 1.4. We select pair of numbers (a, b), such that $1 \leq a \leq 90$ and $1 \leq b \leq 9,000$. We divide a by 10 and take the remainder as i and divide b by 1,000 and take the remainder as R. For a, we see the 2-digited column and for b we see the 4-digited column. We find the following pairs which lead to the selection of the sample.

$$(54,\ 5309) \Rightarrow (3,\ 309)$$
$$(20,\ 2031) \Rightarrow (1,\ 31)$$
$$(83,\ 8341) \Rightarrow (3,\ 341)$$

Since $309 < 356$, $31 < 78$, $341 < 356$, the sample consists of factories with serial numbers $3, 1, 3$.

Suppose we wish to draw a pps sample of 3 factories without replacement. As before, we select the first pair $(53, 5309)$, which leads to the selection of factory No. 3 at the first draw. We now select pair of numbers (a, b) such that $1 \leq a \leq 9$ and $1 \leq b \leq 9,000$, and $a \neq 3$. For a, we see single-digited column and for b, we see 4-digited column. We get the pair $(6, 6133)$ which leads to the selection of factory No. 6. Now we select a pair (a, b), such that $1 \leq a \leq 9$, $1 \leq b \leq 9,000$ and $a \neq 3$ or 6. This pair is $(2, 2031)$. Since $31 < 150$, factory No. 2 is selected. Thus, the pps sample of 3 factories without replacement consists of factories with serial numbers $2, 3, 6$.

EXERCISES

2.1. There are 6 units in a population having sizes $10, 20, 30, 40, 50$ and 60. A sample of 2 units is to be drawn with pps wor. Find the probabilities of inclusion in the sample for (i) each unit and (ii) each pair of units, and hence verify that $\sum_{i=1}^{N} \pi_i = 2$ and $\sum_{j \neq i}^{N} \pi_{ij} = \pi_i$.

2.2. If units are drawn one by one with varying probabilities and without replacement, and at any draw subsequent to the first draw, the probability of selecting a unit from the units available at that draw is proportional to the probability of selecting it at the first draw, show that for samples of size 2, the Yates and Grundy estimator of variance is always positive.

2.3. Suppose that in a sampling system, where the first two units are selected with varying probabilities and without replacement, the probability of selecting a unit at the second draw is proportional to the probability of selecting it at the first draw, while the remaining $(n-2)$ units are selected with equal probability and without replacement from the remaining $(N-2)$ units. Obtain expressions for π_i and π_{ij}. Hence, or otherwise, show that Yates and Grundy estimator of the variance is always positive.

2.4. Show that under Midzuno scheme of sampling

$$T_1 = \frac{\sum_{i=1}^{n} y_i}{\sum_{i=1}^{n} P_i},$$

where the summation is taken over all units in the sample and P_i's are the initial probabilities of selection of the units selected in the sample, is an unbiased estimator of the population total. Obtain an unbiased estimator of the variance of T_1.

2.5. Let $y_1, y_2, \ldots, y_n$ and $P_1, P_2, \ldots, P_n$ be the values of the units in the order in which they are drawn and their initial probabilities of selection. Let $u_i = \frac{y_i}{NP_i}$, $i = 1, 2, \ldots, n$. Show that $\bar{u} = \frac{1}{n}\sum_{i=1}^{n} u_i$ is an unbiased estimator of the population mean. Derive its variance and show that an unbiased estimator of the variance is given by

$$Est.Var(\bar{u}) = \bar{u}^2 - \frac{\sum_{i \neq j}^{n} u'_i u'_j}{n(n-1)},$$

where $u'_1 = \frac{y_1}{NP_1}$, and

$$u'_i = \frac{1}{N}\left[y_1 + y_2 + \ldots + y_{i-1} + \frac{y_i(1-P_1-P_2-\ldots-P_{i-1})}{P_i}\right], i = 2, 3, \ldots, n.$$

2.6. Let there be only two distinct units in a sample of three units selected with ppswr. Show that the estimators

(i) $\frac{1}{3}\left[\frac{y_1}{P_1} + \frac{y_2}{P_2} + \frac{y_1+y_2}{P_1+P_2}\right]$, and

(ii) $\frac{y_1}{1-(1-P_1)^3} + \frac{y_2}{1-(1-P_2)^3}$ are unbiased for the population total Y. If the size measure used for selection is approximately proportional to y, state which of two estimators you would prefer. Give reasons.

2.7. Suppose the population of N units is considered to be derived from the super population with the following model:

$$E(y_i|x_i) = ax_i,$$
$$Var(y_i|x_i) = \sigma_i^2,$$
$$Cov(y_i, y_j|x_i, x_j) = 0, \quad i \neq j.$$

Derive the expected value of the variance of the Horvitz-Thompson estimator of the population total, and show that it reduces to

$$E[Var(\hat{Y}_{HT})] = \sum_{i=1}^{N}\left(\frac{1}{\pi_i} - 1\right)\sigma_i^2,$$

when π_i is made proportional to x_i, $i = 1, 2, \ldots, N$.

2.8. In any sampling design, let π_i be the probability of inclusion of the ith unit in a sample of n units.

(i) Prove that $\hat{Y} = \sum_{i=1}^{n} y_i/\pi_i$, where summation is taken over all the distinct units in the sample, is an unbiased estimator of the population total Y.

(ii) Derive the values of

$$\sum_{i=1}^{N} \pi_i \quad \text{and} \quad \sum_{i=1}^{N}\sum_{j\neq i}^{N} \pi_{ij}.$$

(iii) Obtain $Var(\hat{Y})$.

(iv) Obtain an unbiased estimator of $Var(\hat{Y})$.

2.9. In a sample of n units drawn without replacement, derive the condition for the estimator $\frac{T}{\lambda P(s)}$ to be unbiased for Y, where T is the total of the n sample y values, $P(s)$ is the probability of selecting the sth sample and λ is a constant. Hence, show that the estimator

$$\frac{(y_1 + y_2)(1 - P_1)(1 - P_2)}{(N - 1)P_1P_2(2 - P_1 - P_2)},$$

based on a sample of 2 units drawn with ppswor is unbiased for the population total Y. Further, prove that the above estimator is more efficient than the ordered estimator

$$\frac{(y_1 + y_2)(1 - P_2)}{2(N - 1)P_1P_2}.$$

2.10. Suppose $y_1, y_2, \ldots, y_n$ and $P_1, P_2, \ldots, P_n$ are respectively the values of the sample units and their initial probabilities of selection in order of their selection. Let

$$t_i = y_1 + y_2 + \ldots + y_{i-1} + \frac{y_i}{P_i}(1 - P_1 - P_2 - \ldots - P_{i-1})$$

be the estimator of Y based on the units selected in the i draws. Show that t_i is unbiased for Y. Further consider the estimator

$$\hat{Y} = \frac{1}{n}\sum_{i=1}^{n} t_i.$$

Show that $\hat{Y}$ is an unbiased estimator of Y. Further, show that t_i and t'_i, $i \neq i'$ are uncorrelated. Obtain an unbiased estimator of the variance of $\hat{Y}$.

2.11. Let $n = 2$ in Exercise 2.9. Show that

$$\hat{Y} = \frac{1}{2}\left[\frac{y_1}{P_1}(1 + P_1) + \frac{y_2}{P_2}(1 - P_1)\right]$$

and an unbiased estimator of variance of $\hat{Y}$ is given by

$$Est.Var(\hat{Y}) = \frac{1}{4}(1 - P_1)^2\left(\frac{y_1}{P_1} - \frac{y_2}{P_2}\right)^2.$$

Chapter 3

STRATIFIED SAMPLING

3.1 Stratified Sampling

In stratified sampling, the population is divided into groups called the strata and samples of suitable sizes are selected independently from each stratum. The division of the population into strata is usually done in such a way so that the strata are homogeneous within themselves and exhibit greater variability between them. This sampling is more useful in the following situations:

(i) When estimates are not only required for the population as a whole, but also for groups, forming sub-populations which are considered as strata.
(ii) When the sampling frame is available for sub-regions, then use of stratified sampling is operationally more convenient and more economical, sub-regions being treated as strata.
(iii) When national survey are conducted, stratified sampling is more useful from the point of view of administration and organization of field work.
(iv) In case of skew populations, stratified sampling is more useful since it makes possible to give greater weightage to the few extremely large units for reducing sampling variability.

3.2 Estimation in Stratified Sampling

We assume that a population of N units is divided into k strata of sizes $N_1, N_2, \ldots, N_k$, and that samples from each stratum are selected with SRSWOR. The following notations are followed.

$N(n)$ = the population (sample) size

k = no. of strata

$N_i(n_i) = i^{th}$ stratum population (sample) size

y_{ij} = value of y for the jth unit in the ith stratum

$\bar{Y}_i(\bar{y}_i) = i^{th}$ stratum population (sample) mean defined by

$$N_i\bar{Y}_i = \sum_{j=1}^{N_i} y_{ij}, \quad \text{and} \quad n_i\bar{y}_i = \sum_{j=1}^{n_i} y_{ij}$$

$S_i^2(s_i^2) = i^{th}$ stratum population (sample) mean sum of squares defined by

$$(N_i - 1)S_i^2 = \sum_{j=1}^{N_i}(y_{ij} - \bar{Y}_i)^2, \quad \text{and} \quad (n_i - 1)s_i^2 = \sum_{j=1}^{n_i}(y_{ij} - \bar{y}_i)^2.$$

$w_i = N_i/N,\ i = 1, 2, \ldots, k.$

Theorem 3.2.1. *In stratified sampling,*

(i) $\bar{y}_{st} = \sum_{i=1}^{k} w_i\bar{y}_i$ *is an unbiased estimator of* $\bar{Y}$.

(ii) $Var(\bar{y}_{st}) = \sum_{i=1}^{k} w_i^2 \left(\frac{1}{n_i} - \frac{1}{N_i}\right) S_i^2.$

(iii) $Est.Var(\bar{y}_{st}) = \sum_{i=1}^{k} w_i^2 \left(\frac{1}{n_i} - \frac{1}{N_i}\right) s_i^2.$

Proof. Since $\bar{y}_i$ is the mean of a SRSWOR of n_i units, selected from the ith stratum, it follows that

$$E(\bar{y}_i) = \bar{Y}_i \quad \text{and} \quad Var(\bar{y}_i) = \left(\frac{1}{n_i} - \frac{1}{N_i}\right) S_i^2.$$

Hence,

$$E(\bar{y}_{st}) = \sum_{i=1}^{k} w_i E(\bar{y}_i) = \sum_{i=1}^{k} w_i\bar{Y}_i = \bar{Y}.$$

This proves part (i). Further, since samples from different strata are independently selected, we have

$$Var(\bar{y}_{st}) = \sum_{i=1}^{k} w_i^2 Var(\bar{y}_i) = \sum_{i=1}^{k} w_i^2 \left(\frac{1}{n_i} - \frac{1}{N_i}\right) S_i^2,$$

which proves part (ii) of the theorem.

Lastly, noting that $E(s_i^2) = S_i^2$, we obtain

$$Est.Var(\bar{y}_{st}) = \sum_{i=1}^{k} w_i^2 \left(\frac{1}{n_i} - \frac{1}{N_i}\right) s_i^2.$$

Thus, the theorem is established.

Corollary 3.1. *An unbiased estimator of the population total Y is given by*

$$\hat{Y}_{st} = N\bar{y}_{st}$$

with variance given by

$$Var(\hat{Y}_{st}) = N^2 \sum_{i=1}^{k} w_i^2 \left(\frac{1}{n_i} - \frac{1}{N_i}\right) S_i^2.$$

An unbiased estimator of the $Var(\hat{Y}_{st})$ is obtained as

$$Est.Var(\hat{Y}_{st}) = N^2 \sum_{i=1}^{k} w_i^2 \left(\frac{1}{n_i} - \frac{1}{N_i}\right) s_i^2.$$

3.3 Optimum Allocation

An important question that arises in stratified sampling is: What should be the values of $n_1, n_2, \ldots, n_k$? Optimum allocation is an answer to this question. The problem of optimum allocation is as follows: Determine the value of n_i, $i = 1, 2, \ldots, k$ such that (i) for fixed cost of the survey, the sampling variance is minimized or (ii) for fixed value of sampling variance, the cost of the survey is minimized. The answer to this problem is provided in the following two theorems. The cost function is assumed to be as

$$C = c_0 + \sum_{i=1}^{k} c_i n_i,$$

where c_0 is the overhead cost and c_i is the cost of enumeration and processing the information per unit in the ith stratum.

Theorem 3.3.1. *If the cost of the survey is fixed at C_0, then the sampling variance is minimum if*

$$n_i = \frac{C_0 - c_0}{\sum_{i=1}^{k} w_i S_i \sqrt{c_i}} \cdot \frac{w_i S_i}{\sqrt{c_i}}, \quad i = 1, 2, \ldots, k.$$

Proof. We minimize the sampling variance

$$Var(\bar{y}_{st}) = \sum_{i=1}^{k} w_i^2 \left(\frac{1}{n_i} - \frac{1}{N_i}\right) S_i^2$$

subject to $C_0 = c_0 + \sum_{i=1}^{k} c_i n_i$. We thus minimize

$$\phi = \sum_{i=1}^{k} w_i^2 \left(\frac{1}{n_i} - \frac{1}{N_i}\right) S_i^2 + \lambda(c_0 + \Sigma c_i n_i - C_0)$$

with respect to n_i, λ being Lagrangian multiplier. Equating the partial derivative of ϕ w.r.t. n_i to zero, we obtain

$$n_i = \frac{w_i S_i}{\sqrt{\lambda}\sqrt{c_i}}. \tag{3.3.1}$$

Substituting the value of n_i in $C_0 = c_0 + \sum_{i=1}^{k} c_i n_i$, we obtain

$$\frac{1}{\sqrt{\lambda}} = \frac{C_0 - c_0}{\sum_{i=1}^{k} w_i S_i \sqrt{c_i}}.$$

Substituting the value of $\frac{1}{\sqrt{\lambda}}$ in (3.3.1), we have

$$n_i = \frac{C_0 - c_0}{\sum_{i=1}^{k} w_i \sqrt{c_i} S_i} \cdot \frac{w_i S_i}{\sqrt{c_i}}, \tag{3.3.2}$$

which proves the theorem.

Remark. Substituting the value of n_i from (3.3.2) in $Var(\bar{y}_{st})$, we obtain for fixed cost C_0, the minimum value of $Var(\bar{y}_{st})$ as

$$Var(\bar{y}_{st})_{min} = \frac{\left(\sum_{i=1}^{k} w_i S_i c_i\right)^2}{C_0 - c_0} - \frac{1}{N}\sum_{i=1}^{k} N_i S_i^2. \tag{3.3.3}$$

Theorem 3.3.2. *If the sampling variance is fixed at V_0, then cost of the survey is minimum if*

$$n_i = \frac{\sum_{i=1}^{k} w_i\sqrt{c_i}S_i}{V_0 + \frac{1}{N}\sum_{i=1}^{k} w_i S_i^2} \cdot \frac{w_i S_i}{\sqrt{c_i}}, \quad i = 1, 2, \ldots, k.$$

Proof. We minimize the cost of survey $C = c_0 + \sum_{i=1}^{k} c_i n_i$, subject to

$$V_0 = \sum_{i=1}^{k} w_i^2 \left(\frac{1}{n_i} - \frac{1}{N_i}\right) S_i^2.$$

We thus minimize

$$\phi = c_0 + \Sigma c_i n_i + \lambda \left(\sum_{i=1}^{k} w_i^2 \left(\frac{1}{n_i} - \frac{1}{N_i}\right) S_i^2 - V_0\right)$$

with respect to n_i, λ being the Lagrangian multiplier. Equating the partial derivative of ϕ w.r.t. n_i to zero, we obtain

$$n_i = \sqrt{\lambda}\frac{w_i S_i}{\sqrt{c_i}}. \tag{3.3.4}$$

Substituting this value of n_i in $V_0 = \sum_{i=1}^{k} w_i^2 \left(\frac{1}{n_i} - \frac{1}{N_i}\right) S_i^2$, we obtain

$$\sqrt{\lambda} = \frac{\sum_{i=1}^{k} w_i\sqrt{c_i}S_i}{V_0 + \frac{1}{N}\sum_{i=1}^{k} w_i S_i^2}.$$

Substituting this value of $\sqrt{\lambda}$ in (3.3.4), we obtain n_i as

$$n_i = \frac{\sum_{i=1}^{k} w_i\sqrt{c_i}S_i}{V_0 + \frac{1}{N}\sum_{i=1}^{k} w_i S_i^2} \cdot \frac{w_i S_i}{\sqrt{c_i}}, \quad i = 1, 2, \ldots, k. \tag{3.3.5}$$

Remark 1. Substituting the value of n_i from (3.3.5) in $C = c_0 + \sum_{i=1}^{k} c_i n_i$, we obtain for fixed value V_0 of the sampling variance, the minimum value of the cost of the survey as

$$C_{min} = c_0 + \frac{\left(\sum_{i=1}^{k} w_i \sqrt{c_i} S_i\right)^2}{V_0 + \frac{1}{N}\sum_{i=1}^{k} N_i S_i^2}. \tag{3.3.6}$$

Remark 2. From (3.3.1) and (3.3.4), it follows that when either cost is fixed or variance is fixed, the optimum value of n_i is proportional to $\frac{w_i S_i}{\sqrt{c_i}}$.

Theorem 3.3.3. (Neyman allocation): *If the total sample size n is fixed, then the sampling variance is minimum if*

$$n_i = n \frac{w_i S_i}{\sum_{i=1}^{k} w_i S_i}.$$

Proof. We minimize

$$\phi = \sum_{i=1}^{k} w_i^2 \left(\frac{1}{n_i} - \frac{1}{N_i}\right) S_i^2 + \lambda \left(\sum_{i=1}^{k} n_i - n\right).$$

Equating the partial derivative of ϕ w.r.t. n_i to zero, we get

$$n_i = \frac{1}{\sqrt{\lambda}} w_i S_i.$$

Substituting this value of n_i in $n = \sum_{i=1}^{k} n_i$, we obtain

$$\frac{1}{\sqrt{\lambda}} = \frac{n}{\sum_{i=1}^{k} w_i S_i}.$$

Substituting this value of $\frac{1}{\sqrt{\lambda}}$ in (3.3.7), we get

$$n_i = n \cdot \frac{w_i S_i}{\sum_{i=1}^{k} w_i S_i}, \quad i = 1, 2, \ldots, k. \tag{3.3.8}$$

The value of the minimum variance in this case is given by

$$Var(\bar{y}_{st})_{Ney} = \frac{1}{n}\left(\sum_{i=1}^{k} w_i S_i\right)^2 - \frac{1}{N}\sum_{i=1}^{k} w_i S_i^2. \tag{3.3.9}$$

3.4 Proportional Allocation

When no informatin about fixed cost or desired sampling variance is available, the allocation of n_i may be done in proportion to N_i. So, in proportional allocation, we take

$$n_i = nw_i, \quad i = 1, 2, \ldots, k. \tag{3.4.1}$$

The sampling variance under proportional allocation is obtained as

$$Var(\bar{y}_{st})_{prop} = \left(\frac{1}{n} - \frac{1}{N}\right) \sum_{i=1}^{k} w_i S_i^2. \tag{3.4.2}$$

This allocation was originally suggested by Bowley (1926).

If the cost of the survey is fixed at C_0 and the cost function is assumed to be $C = c_0 + \sum_{i=1}^{k} c_i n_i$, then, the total sample size n under proportional allocation is given by

$$n = \frac{C_0 - c_0}{\sum_{i=1}^{k} c_i w_i}. \tag{3.4.3}$$

3.5 Comparison of Stratified Sampling with SRSWOR

We shall compare stratified sampling (i) under proportional allocation, (ii) under Neyman allocation and (iii) under arbitrary allocation, with SRSWOR.

(i) Under Proportional Allocation: Here, from (3.4.2), we find that the variance of $\bar{y}_{st}$ is given by

$$Var(\bar{y}_{st})_{prop} = \left(\frac{1}{n} - \frac{1}{N}\right) \sum_{i=1}^{k} w_i S_i^2. \tag{3.5.1}$$

Now, variance of $\bar{y}$ in the case of SRSWOR is given by

$$Var(\bar{y})_{SRS} = \left(\frac{1}{n} - \frac{1}{N}\right) S^2 \tag{3.5.2}$$

where we need to express S^2 in terms of S_i^2. For this, we note that

$$(N-1)S^2 = \sum_{i=1}^{k} (N_i - 1) S_i^2 + \sum_{i=1}^{k} N_i (\bar{Y}_i - \bar{Y})^2. \tag{3.5.3}$$

Dividing (3.5.3) by N and assuming N very large so that $\frac{1}{N}$ is negligible, we obtain

$$S^2 = \sum_{i=1}^{k} w_i S_i^2 + \sum_{i=1}^{k} w_i(\bar{y}_i - \bar{Y})^2. \tag{3.5.4}$$

Hence (3.5.2) becomes

$$Var(\bar{y})_{SRS} = \frac{1}{n}\left[\sum_{i=1}^{k} w_i S_i^2 + \sum_{i=1}^{k} w_i(\bar{Y}_i - \bar{Y})^2\right] \tag{3.5.5}$$

since $\frac{1}{N}$ is assumed to be negligible. Also, under the assumption of $\frac{1}{N}$ being negligible, (3.5.1) becomes

$$Var(\bar{y})_{prop} = \frac{1}{n}\sum_{i=1}^{k} w_i S_i^2. \tag{3.5.6}$$

From (3.5.5) and (3.5.6), we get

$$Var(\bar{y})_{SRS} - Var(\bar{y}_{st})_{prop} = \frac{1}{n}\sum_{i=1}^{k} w_i(\bar{Y}_i - \bar{Y})^2 \geq 0. \tag{3.5.7}$$

Thus, we see that stratified sampling under proportional allocation is more efficient than SRSWOR. Further, we note that reduction in variance is more if strata means differ more.

(ii) Under Neyman Allocation: When $\frac{1}{N}$ is assumed to be negligible, from (3.3.9) we obtain the variance of $\bar{y}_{st}$ under Neyman allocation as

$$Var(\bar{y}_{st})_{Ney} = \frac{1}{n}\left(\sum_{i=1}^{k} w_i S_i\right)^2. \tag{3.5.8}$$

From (3.5.5) and (3.5.8), one obtains

$$Var(\bar{y})_{SRS} - Var(\bar{y}_{st})_{Ney} = \frac{1}{n}\sum_{i=1}^{k} w_i(\bar{Y}_i - \bar{Y})^2 + \frac{1}{n}\sum_{i=1}^{k} w_i(S_i - \bar{S})^2 \geq 0, \tag{3.5.9}$$

where $\bar{S} = \sum_{i=1}^{k} w_i S_i$. From (3.5.9), we see that stratified sampling is more efficient than SRSWOR and reduction in variance is more if strata means and strata standard derivations differ more.

Further, from (3.5.6) and (3.5.8) one finds that

$$Var(\bar{y}_{st})_{prop} - Var(\bar{y}_{st})_{Ney} = \sum_{i=1}^{k} w_i(S_i - \bar{S})^2 \geq 0. \tag{3.5.10}$$

Hence, stratified sampling under Neyman allocation is more efficient than that under proportional allocation.

Also, from (3.5.7) and (3.5.10) one obtains

$$Var(\bar{y})_{SRS} \geq Var(\bar{y}_{st})_{prop} \geq Var(\bar{y}_{st})_{Ney}. \tag{3.5.11}$$

(iii) Under Arbitrary Allocation: The variance of $\bar{y}_{st}$ in stratified sampling has been obtained in Theorem 3.2.1 as

$$\begin{aligned} Var(\bar{y}_{st}) &= \sum_{i=1}^{k} w_i^2 \left(\frac{1}{n_i} - \frac{1}{N_i} \right) S_i^2 \\ &= \sum_{i=1}^{k} \frac{1}{n_i} w_i^2 S_i^2 - \frac{1}{N} \sum_{i=1}^{k} w_i S_i^2. \end{aligned}$$

Since we assume $\frac{1}{N}$ negligible, the above variance can be written as

$$Var(\bar{y}_{st}) = \sum_{i=1}^{k} w_i^2 S_i^2 / n_i. \tag{3.5.12}$$

So, from (3.5.5) and (3.5.12), we obtain

$$Var(\bar{y})_{SRS} - Var(\bar{y}_{st}) = \sum_{i=1}^{k} \left(\frac{1}{n} - \frac{w_i}{n_i} \right) w_i S_i^2 + \frac{1}{n} \sum_{i=1}^{k} w_i (\bar{Y}_i - \bar{Y})^2. \tag{3.5.13}$$

The second term on the r.h.s. of (3.5.13) is non-negative, but the first term on the r.h.s. of (3.5.13) could be > 0, $= 0$ or < 0. Hence, in general, we cannot say stratified sampling is always more efficient than SRSWOR. However, we note that if $\frac{n_i}{n} \geq w_i$, $i = 1, 2, \ldots, k$, then stratified samplling is more efficient than SRSWOR.

3.6 Estimation of Gain in Efficiency from Stratified Sample

From Theorem 3.2.1, we see that an unbiased estimator of $Var(\bar{y}_{st})$ is given by

$$Est.Var(\bar{y}_{st}) = \sum_{i=1}^{k} \left(\frac{1}{n_i} - \frac{1}{N_i} \right) w_i^2 s_i^2. \tag{3.6.1}$$

Now, we shall obtain an unbiased estimator of $Var(\bar{y})_{SRS}$ on the basis of stratified sample. We know that

$$Var(\bar{y})_{SRS} = \left(\frac{1}{n} - \frac{1}{N} \right) S^2. \tag{3.6.2}$$

Further, it is easily verified that

$$(N-1)S^2 = \sum_{i=1}^{k}(N_i-1)S_i^2 + \sum_{i=1}^{k} N_i(\bar{Y}_i - \bar{Y})^2. \qquad (3.6.3)$$

Now,

$$Var(\bar{y}_i) = \left(\frac{1}{n_i} - \frac{1}{N_i}\right) S_i^2 = E(\bar{y}_i^2) - \bar{Y}_i^2.$$

Hence, an unbiased estimator of $\bar{Y}_i^2$ is obtained as

$$Est.(\bar{Y}_i^2) = \bar{y}_i^2 - \left(\frac{1}{n_i} - \frac{1}{N_i}\right) s_i^2. \qquad (3.6.4)$$

Further,

$$\begin{aligned} Var(\bar{y}_{st}) &= \sum_{i=1}^{k}\left(\frac{1}{n_i} - \frac{1}{N_i}\right) w_i^2 S_i^2 \\ &= E(\bar{y}_{st}^2) - \bar{Y}^2, \end{aligned}$$

from which one obtains an unbiased estimator of $\bar{Y}^2$ as

$$Est.(\bar{Y}^2) = \bar{y}_{st}^2 - \sum_{i=1}^{k}\left(\frac{1}{n_i} - \frac{1}{N_i}\right) w_i^2 s_i^2. \qquad (3.6.5)$$

Using (3.6.4) and (3.6.5), we obtain an unbiased estimator of

$$\sum_{i=1}^{k} N_i(\bar{Y}_i - \bar{Y})^2 = \sum_{i=1}^{k} N_i \bar{Y}_i^2 - N\bar{Y}^2$$

as

$$Est. \sum_{i=1}^{k} N_i(\bar{Y}_i - \bar{Y})^2 = \sum_{i=1}^{k} N_i \bar{y}_i^2 - N\bar{y}_{st}^2 - N\sum_{i=1}^{k}\left(\frac{1}{n_i} - \frac{1}{N_i}\right) w_i(1-w_i)s_i^2. \qquad (3.6.6)$$

Clearly,

$$Est.\left(\sum_{i=1}^{k}(N_i-1)S_i^2\right) = \sum_{i=1}^{k}(N_i-1)s_i^2. \qquad (3.6.7)$$

Hence, using (3.6.7), (3.6.6), and (3.6.3), we obtain an unbiased estimator of S^2 as

$$\begin{aligned} Est.S^2 = \frac{1}{N-1}\Bigg[&\sum_{i=1}^{k}(N_i-1)s_i^2 + \sum_{i=1}^{k} N_i \bar{y}_i^2 \\ &- N\bar{y}_{st}^2 - N\sum_{i=1}^{k}\left(\frac{1}{n_i} - \frac{1}{N_i}\right) w_i(1-w_i)s_i^2\Bigg]. \end{aligned} \qquad (3.6.8)$$

Therefore,

$$Est.Var(\bar{y})_{SRS} = \left(\frac{1}{n} - \frac{1}{N}\right) \cdot \frac{1}{N-1}\left[\sum_{i=1}^{k}(N_i - 1)s_i^2 + \sum_{i=1}^{k} N_i\bar{y}_i^2 - N\bar{y}_{st}^2\right]. \tag{3.6.9}$$

Then, efficiency E of stratified sampling relative to simple random sample without replacement is estimated by

$$\hat{E} = \frac{Est.Var(\bar{y})_{SRS}}{Est.Var(\bar{y}_{st})} \times 100\%. \tag{3.6.10}$$

Then, gain in efficiency is estimated by $(\hat{E} - 100)\%$.

3.7 Estimation of a Proportion in Stratified Sampling

The units of the population are divided into two classes C and $\bar{C}$, C consisting of units having a certain attribute and $\bar{C}$ consisting of units not having that attribute. Let M be the number of units in the class C and $N-M$ be the number of units in $\bar{C}$. Then $P = \frac{M}{N}$ is the proportion of units having the particular attribute. Let $Q = 1 - P$. Suppose that in the ith stratum, there are M_i units possessing the given attribute, $i = 1, 2, \ldots, k$. Let $P_i = \frac{M_i}{N_i}$ be the proportion of units having the particular attribute in the ith stratum, $i = 1, 2, \ldots, k$ and let $Q_i = 1 - P_i$. We now define y_{ij}, $i = 1, 2, \ldots, k_j$, $j = 1, 2, \ldots, N_i$ as follows.

$$\begin{aligned} y_{ij} &= 1, \quad \text{if the } j^{th} \text{ unit in the } i^{th} \text{ stratum belongs to class } C \\ &= 0, \quad \text{if it does not belong to class } C. \end{aligned}$$

Then, clearly, we have the following results.

$$\bar{Y}_i = P_i,$$

$$S_i^2 = \frac{N_i P_i Q_i}{(N_i - 1)}, \quad i = 1, 2, \ldots, k,$$

and

$$P = \sum_{i=1}^{k} w_i P_i.$$

We now consider a stratified sample $(n_1, n_2, \ldots, n_k)$. Let m_i be the number of units in n_i, possessing the given attribute, and $p_i = \frac{m_i}{n_i}$, $q_i = 1 - p_i$, $i = 1, 2, \ldots, k$. Then, one obtains

$$\bar{y}_i = p_i, \quad s_i^2 = \frac{n_i p_i q_i}{(n_i - 1)}.$$

Then, using Theorem 3.2.1, we obtain

Theorem 3.7.1.

(i) $p_{st} = \sum_{i=1}^{k} w_i p_i$ *is an unbiased estimator of* P.

(ii) $Var(p_{st}) = \sum_{i=1}^{k} \left(\frac{1}{n_i} - \frac{1}{N_i}\right) w_i^2 \cdot \frac{N_i P_i Q_i}{(N_i - 1)}$.

(iii) *An unbiased estimator of* $Var(p_{st})$ is given by

$$Est.Var(p_{st}) = \sum_{i=1}^{k} \left(\frac{1}{n_i} - \frac{1}{N_i}\right) w_i^2 \frac{n_i p_i q_i}{n_i - 1}.$$

Optimum Allocation:

(i) If C_0, the cost of the survey is fixed, then variance is minimum if

$$n_i = \frac{(C_0 - c_0)}{\sum_{i=1}^{k} w_i \sqrt{\frac{c_i N_i P_i Q_i}{N_i - 1}}} \cdot \frac{w_i \sqrt{N_i P_i Q_i}}{\sqrt{c_i (N_i - 1)}}, \quad i = 1, 2, \ldots, k. \qquad (3.7.1)$$

This follows from Theorem 3.3.1.
If $\frac{N_i}{(N_i - 1)} \approx 1$, then $n_i = \frac{C_0 - c_0}{\sum_{i=1}^{k} w_i \sqrt{c_i P_i Q_i}} \cdot \frac{w_i \sqrt{P_i Q_i}}{\sqrt{c_i}}$.

(ii) If the sampling variance is fixed at V_0, then the cost of the survey is minimum if

$$n_i = \frac{\sum_{i=1}^{k} w_i \sqrt{\frac{c_i N_i P_i Q_i}{(N_i - 1)}}}{V_0 + \frac{1}{N} \sum_{i=1}^{k} w_i \frac{N_i P_i Q_i}{N_i - 1}} \cdot \frac{w_i \sqrt{N_i P_i Q_i}}{\sqrt{c_i (N_i - 1)}}. \qquad (3.7.2)$$

This follows from Theorem 3.3.2.

If $\frac{N_i}{(N_i-1)} \approx 1$, then

$$n_i = \frac{\sum_{i=1}^{k} w_i\sqrt{c_iP_iQ_i}}{V_0 + \frac{1}{N}\sum_{i=1}^{k} w_iP_iQ_i} \cdot \frac{w_i\sqrt{P_iQ_i}}{\sqrt{c_i}}, \quad i = 1, 2, \ldots, k.$$

(iii) Neyman Allocation: If the total sample size n is fixed then the sampling variance is minimum if

$$n_i = n \cdot \frac{w_i\sqrt{\frac{N_iP_iQ_i}{(N_i-1)}}}{\sum_{i=1}^{k} w_i\sqrt{\frac{N_iP_iQ_i}{(N_i-1)}}}, \quad i = 1, 2, \ldots, k. \tag{3.7.3}$$

This follows from Theorem 3.3.3.
If $\frac{N_i}{(N_i-1)} \approx 1$, then

$$n_i = n\frac{w_i\sqrt{P_iQ_i}}{\sum_{i=1}^{k} w_i\sqrt{P_iQ_i}}, \quad i = 1, 2, \ldots, k.$$

From (3.4.2), it follows that in the case of proportional allocation,

$$Var(p_{st})_{prop} = \left(\frac{1}{n} - \frac{1}{N}\right)\sum_{i=1}^{k} \frac{w_iN_iP_iQ_i}{(N_i-1)}, \tag{3.7.4}$$

which becomes

$$Var(p_{st})_{prop} = \left(\frac{1}{n} - \frac{1}{N}\right)\sum_{i=1}^{k} w_iP_iQ_i,$$

if $\frac{N_i}{(N_i-1)} \approx 1$.

Comparison with SRSWOR

From (3.5.7), we obtain that

$$Var(p_{st})_{SRS} - Var(p_{st})_{prop} = \frac{1}{n}\sum_{i=1}^{k} w_i(P_i - P)^2 \geq 0,$$

which shows that stratified sampling with proportional allocation is more efficient than SRSWOR.

From (3.5.9), one obtains that

$$Var(p)_{SRS} - Var(p_{st})_{Ney} = \frac{1}{n}\sum_{i=1}^{k} w_i(P_i - P)^2 + \frac{1}{n}\sum_{i=1}^{k} w_i \cdot \left[\sqrt{\frac{N_iP_iQ_i}{N_i-1}} - \sum_{i=1}^{k} w_i\sqrt{\frac{N_iP_iQ_i}{N_i-1}}\right]^2. \tag{3.7.5}$$

If $\frac{N_i}{(N_i-1)} \approx 1$, then (3.7.5) reduces to

$$Var(p)_{SRS} - Var(p_{st})_{Ney} = \frac{1}{n}\sum_{i=1}^{k} w_i(P_i - P)^2 + \frac{1}{n}\sum_{i=1}^{k} w_i \cdot \left[\sqrt{P_iQ_i} - \sum_{i=1}^{k} w_iP_iQ_i\right]^2. \tag{3.7.6}$$

Estimation of Gain in Efficiency

For N large, we derive the expression for estimation of gain in efficiency due to stratification.

From (3.6.1), when N is large, we have

$$Est.Var(p_{st}) = \sum_{i=1}^{k} \frac{1}{(n_i-1)} w_i^2 p_i q_i. \tag{3.7.7}$$

From (3.6.9), when N is large, we have

$$Est.Var(p)_{SRS} = \frac{1}{n}\left[\sum_{i=1}^{k} \frac{w_i n_i p_i q_i}{(n_i-1)} + \sum_{i=1}^{k} w_i^2 p_i^2 - \bar{p}_{st}^2 - \sum_{i=1}^{k} w_i(1-w_i)p_iq_i(n_i-1)\right]. \tag{3.7.8}$$

Hence, estimate of efficiency is obtained as

$$\hat{E} = \frac{Est.Var(p)_{SRS}}{Est.Var(p_{st})} \times 100\%$$

and the estimate of gain in efficiency is $(\hat{E} - 100)\%$.

EXERCISES

3.1. Suppose in stratified sampling with SRSWOR, the total sample size of n was allocated by mistake in proportion to $N_iS_i^2$ instead of usual optimum allocation proportional to N_iS_i. How does allocation compare with proportional and optimum allocation?

3.2. Suppose it is desired to estimate the difference between the proportiona of a certain attribute in two villages, one a model village having N_1 persons and the other village having N_2 persons. Let P_1 and P_2 be the proportions of persons having the specified attribute in the two villages and c_1 and c_2 be the costs per person of collecting the data in the two villages respectively. Assuming the cost to be fixed at C_0, determine the optimum allocation of the total sample size to the two villages when SRSWR is used in each village.

3.3. Let t_{i1} and t_{i2} be unbiased estimators of the ith stratum ($i = 1, 2, \ldots, k$) total based on two independent samples. Show that

(i) $V_1 = \sum_{i=1}^{k} \frac{(t_{i1} - t_{i2})^2}{4}$.

(ii) $V_2 = \frac{\left(\sum_{i=1}^{k} t_{i1} - \sum_{i=1}^{k} t_{i2}\right)^2}{4}$ are unbiased estimators of the variance of the combined estimator $\frac{\sum_{i=1}^{k}(t_{i1} + t_{i2})}{2}$.

3.4. A population of N units is divided into two strata of sizes N_1 and N_2 units and samples of sizes n_1 and n_2 units are selected from each of the two strata with SRSWR. Show that the efficiency of the estimator of $\bar{Y}$ in this case as compared with that of optimum allocation of the total sample size $n = (n_1+n_2)$ is not less than $\frac{4t}{(1+t)^2}$, where $t = \frac{n_1 n_2'}{n_2 n_1'}$, where n_1' and n_2' are the optimum allocation sizes in the two strata.

3.5. A population of N units is divided into k strata at random and the ith stratum has a specified number of units N_i, $i = 1, 2, \ldots, k$. From each stratum, a SRSWOR is selected, using proportional allocation. Show that the variance of the estimator of the overall mean is equal to that of the sample mean of an unstratified SRSWOR of the same overall sample size.

3.6. In stratified sampling, suppose that the number of units in a stratum is the same for all strata. A stratified sample of overall size n is selected with SRSWR and with equal allocation. Show that the sampling variance of the estimator of $\bar{Y}$ can be expressed as

$$V_{st} = V_{SRS} - \frac{1}{kn}\sum_{i=1}^{k}(\bar{Y}_i - \bar{Y})^2$$

where V_{SRS} = the variance of an unstratified SRSWR in estimating $\bar{Y}$.

3.7. In stratified sampling where one unit is selected from each stratum, the sampling variance is usuallly estimated by adopting the method of collapsed strata. This method consists of pairing the strata to form collapsed strata and estimating the sampling variance as if two units were selected from each collapsed stratum.

Assuming the proportion of units w_i is the same for each of the two strata forming the ith pair, $i = 1, 2, \ldots, \frac{k}{2}$ and assuming SRSWOR within strata, show that the variance estimator

$$v_1 = Est.Var(\hat{\bar{Y}}_{st}) = \sum_{i=1}^{\frac{k}{2}} w_i^2(y_{i1} - y_{i2})^2,$$

where y_{i1}, y_{i2} are the values of units selected from the two strata forming the ith collapsed stratum, overestimates $Var(\hat{\bar{Y}}_{st})$ and that the bias is small when the two strata forming the ith collapsed stratum have approximately the same mean, $i = 1, 2, \ldots, \frac{k}{2}$.

3.8. In Exercise 3.7, if the proportions of units w_{i1}, w_{i2} for the two strata forming the ith collapsed stratum are not the same for all i, then consider the variance estimator

$$v = Est.Var(\hat{\bar{Y}}_{st}) = \sum_{i=1}^{\frac{k}{2}}[w_{i1}^2 y_{i1} - w_{i2}^2 y_{i2}](y_{i1} - y_{i2})$$

and obtain its bias.

3.9. A sample of n units is drawn with SRSWOR and two post-strata are formed at the estimation stage. There are two possibilities: (i) each post-stratum contains at least one sample unit and (ii) one of the two post-strata is empty, that is, contains no sample unit. Consider the estimators

(a) $\hat{\bar{Y}}_1 = w_1\bar{y}_1 + w_2\bar{y}_2$,

(b) $\hat{\bar{Y}}_2 = \alpha D_1\bar{y}_1 + (1-\alpha)D_2\bar{y}_2$,

where w_1, $w_2 = 1 - w_1$ are proportions of units, and $\bar{y}_1$, $\bar{y}_2$ are the sample means for the two post-strata means, and α is 1 or 0 according as stratum 2 or 1 is empty, and $D_1 = \frac{w_1}{P_1}$, $D_2 = \frac{w_2}{P_2}$, $P_2 = 1 - P_1$, P_1

being the conditional probability that the stratum 2 is empty given that possibility (ii) has occurred. Show that the estimators (a) and (b) are conditionally unbiased given that possibility (i) or (ii) has occurred respectively. Find the bias and variance of

$$\hat{\hat{Y}} = \lambda\hat{\hat{Y}}_1 + (1-\lambda)\hat{\hat{Y}}_2,$$

where λ is 1 or 0 accordingly as possibilitiy (i) or (ii) has occurred.

3.10. A population consists of k strata of sizes N_i and mean values $\bar{Y}_i$, $i = 1, 2, \ldots, k$. It is required to estimate r linear functions $L_s = \sum_{i=1}^{k} \ell_{is}\bar{Y}_i$, $s = 1, 2, \ldots, r$ of strata means by selecting n_i units from the ith stratum with SRSWOR. Assuming the cost function to be $C = \sum_{i=1}^{k} c_i n_i^g$, $g > 0$, obtain the values of n_i such that for a fixed cost, the expected loss given by $\sum_{s=1}^{r} \mu_s Var(\hat{L}_s)$ is minimized, $\hat{L}_s$ being the estimator of L_s and μ_s and ℓ_{is} are known constants.

3.11. A random sample of size n is selected from a population and the sample units are allocated to k strata on the basis of information collected about them. Denoting by n_i (a random variable) the number of sample units falling in stratum i, show that the variance of $\sum_{i=1}^{k} w_i\bar{y}_i$, ($w_i$ known) is given by

$$\left(\frac{1}{n} - \frac{1}{N}\right)\sum_{i=1}^{k} w_i S_i^2 + \frac{1}{n^2}\sum_{i=1}^{k}(1-w_i)S_i^2.$$

Compare this variance with the one obtained in stratified random sampling with proportional allocation when it is feasible to select units within strata.

3.12. A population is divided into k strata, N_i being the number of units in the stratum i, $i = 1, 2, \ldots, k$. Information on an auxiliary character x is known for each unit. A sample s_n of size n is selected from the entire population such that the probability that s_n be selected is proportional to $\left(\sum_{i=1}^{k} N_i\bar{x}_i\right)_{s_n}$, where $\bar{x}_i$ is the mean of x based on the n_i sample units taken from the stratum i. Prove that $\hat{Y}_{st} =$

$X\dfrac{\sum_{i=1}^{k} N_i\bar{y}_i}{\sum_{i=1}^{k} N_i\bar{x}_i}$ is an unbiased estimator of the population total Y and obtain its variance. Also, obtain an unbiased estimator of $Var(\hat{Y}_{st})$.

3.13. The households in a town are to be sampled in order to estimate the average amount of assets per household that are readily convertible in cash. The households are stratified into a high-rent and a low-rent stratum. A house in the high-rent stratum is thought to have about nine times as much assets as one in the low-rent stratum and S_i is expected to be proportional to the square root of the stratum mean.

There are 4,000 households in the high-rent stratum and 20,000 in the low-rent stratum. (i) How would you distribute a sample of 1,000 households between the two strata? (ii) If the object is to estimate the difference between assets per household in the two strata, how should the sample be distributed over two strata?

3.14. Show that

$$\frac{Var(\bar{y}_{st})_{prop}}{Var(\bar{y}_{st})_{opt}} = \frac{w_1c_1 + w_2c_2}{(w_1\sqrt{c_1} + w_2\sqrt{c_2})^2}.$$

If $w_1 = w_2$, show that the relative increase in variance from using proportional allocation is given by

$$\frac{(\sqrt{c_1} - \sqrt{c_2})^2}{(\sqrt{c_1} + \sqrt{c_2})^2}.$$

A population is divided into two strata, and a sample of size n is selected with $n_1 = n_2$ instead of the values given by the Neyman allocation. If $Var(\bar{y}_{st})$ and $Var(\bar{y}_{st})_{opt}$ denote the variance given by $n_1 = n_2$ and the Neyman allocation, respectively, show that the relative increase in variance is

$$\frac{Var(\bar{y}_{st}) - Var(\bar{y}_{st})_{opt}}{Var(\bar{y}_{st})_{opt}} = \left(\frac{\lambda - 1}{\lambda + 1}\right)^2,$$

where $\lambda = \frac{n_1}{n_2}$ as given by Neyman allocation (ignore fpc).

3.15 Let the cost function be represented by $C = c_0 + \sum_{i=1}^{K} c_i\sqrt{n_i}$, where c_0 and c_i are known numbers. Show that in order to minimize $Var(\bar{y}_{st})$ for fixed total cost, n_i must be proportional to

$$\left(\frac{w_i^2 S_i^2}{c_i}\right)^{\frac{2}{3}}.$$

Determine n_i for a sample of 1,000 units under the following conditions:

Stratum	w_i	S_i	c_i
1	0.4	4	1
2	0.3	5	2
3	0.2	6	4

3.16. Let $V(\bar{y}_{st})_{prop}$ be the variance of the mean of a stratified random sample of size n under proportional allocation and $Var(\bar{y})_{SRS}$ be the variance of the mean of a SRSWOR of size n. Show that the ratio

$$\frac{Var(\bar{y}_{st})_{prop}}{Var(\bar{y})_{SRS}}$$

does not depend upon the size of the sample and the ratio

$$\frac{Var(\bar{y}_{st})_{min}}{Var(\bar{y}_{st})_{prop}}$$

decreases as n increases.

Chapter 4

SYSTEMATIC SAMPLING

4.1 Systematic Sampling

Another sampling technique which is operationally more convenient than simple random sampling and which at the same time ensures equal probability of inclusion for every unit in the sample is *Systematic Sampling.* In systematic samplinig the first unit is selected at random and the remaining units are selected according to some pre-determined pattern or systematically. For example, suppose the number of units in the population is $N = nk$, where n is the sample size. A random number from 1 to k is selected at random. Suppose it to be i. Then, the first unit in the systematic sample is selected as the unit whose serial number is i, and the other units in the systematic sample are selected by selecting every k^{th} unit after the ith unit. Thus, the selection of the first unit is done at random, while the other units are selected systematically. The above systematic sampling is called every k^{th} systematic sample, k being called the sampling interval. This is also known as linear systematic sampling.

EXAMPLE: Let $N = 100$ and $n = 10$. Hence $k = 10$. Suppose the random number selected from 1 to 10 is 7. Then the systematic sample consists of units with serial numbers

$$7, 17, 27, 37, 47, 57, 67, 77, 87, 97.$$

Systematic sampling is similar to stratified sampling in which the population is divided into n strata of k units each and one unit is drawn from each stratum. But the difference between systematic sampling and stratified sampling is that in systematic sampling, the relative positions of units selected from the strata remains the same within the strata, while in stratified sampling the positions of units selected within the strata are determined in a random manner.

In what follows we shall assume that $N = nk$. At the end, we shall discuss the type of sampling when $N \neq nk$.

4.2 Estimation in Systematic Sampling

Let there be $N = nk$ units in the population. A systematic sample of size n is selected by selecting a unit with serial number from 1 to k at random. Hence, there will be k systematic samples, each having probability $\frac{1}{k}$ of being selected as the one particular sample. Let y be the study variable, and y_{ij} denote the value of y associated with the jth unit in the ith systematic sample, $i = 1, 2, \ldots, k$ and $j = 1, 2, \ldots, n$. Let $\bar{y}_i = \frac{1}{n}\sum_{j=1}^{n} y_{ij}$ be the ith systematic sample mean. Let $\bar{Y} = \frac{1}{N}\sum_{i=1}^{k}\sum_{j=1}^{n} y_{ij}$ be the population mean. The k systematic samples, together with their means are given in Table 4.1.

Let $\bar{y}_{..} = \frac{1}{k}\sum\limits_{i=1}^{k} \bar{y}_{i\cdot}$. We note here that $\bar{Y} = \bar{y}_{..}$.

Let $\bar{y}_{sy}$ denote the mean of the selected systematic sample. Clearly, $\bar{y}_{sy}$ assumes values $\bar{y}_{1\cdot}, \bar{y}_{2\cdot}, \ldots, \bar{y}_{k\cdot}$ with probability $\frac{1}{k}$. Hence we get

$$E(\bar{y}_{sy}) = \sum_{i=1}^{k} \frac{1}{k}\bar{y}_{i\cdot} = \bar{y}_{..} = \bar{Y}. \qquad (4.2.1)$$

Table 4.1 Systematic Samples of n units from a population of $N = nk$ units

Systematic Sample No.	Sample Composition 1 2 ... n	Prob.	Sample mean
1	$y_{11}, y_{12}, \ldots, y_{1n}$	$\frac{1}{k}$	$\bar{y}_1$
2	$y_{21}, y_{22}, \ldots, y_{2n}$	$\frac{1}{k}$	$\bar{y}_2$
.			
.			
.			
i	$y_{i1}, y_{i2}, \ldots, y_{in}$	$\frac{1}{k}$	$\bar{y}_i$
.			
.			
.			
k	$y_{k1}, y_{k2}, \ldots, y_{kn}$	$\frac{1}{k}$	$\bar{y}_k$

Further,

$$Var(\bar{y}_{sy}) = \sum_{i=1}^{k} \frac{1}{k}(\bar{y}_{i\cdot} - \bar{y}_{\cdot\cdot})^2 = \frac{1}{k}\sum_{i=1}^{k}(\bar{y}_{i\cdot} - \bar{y}_{\cdot\cdot})^2. \tag{4.2.2}$$

Thus. we have proved the following theorem.

Theorem 4.2.1. *In systematic sampling,*

(i) *the systematic sample mean $\bar{y}_{sy}$ is an unbiased estimator of population mean $\bar{y}_{\cdot\cdot}$.*

(ii) $Var(\bar{y}_{sy}) = \frac{1}{k}\sum_{i=1}^{k}(\bar{y}_{i\cdot} - \bar{y}_{\cdot\cdot})^2$.

4.3 Comparison with SRSWOR

For comparison with SRSWOR, we shall derive alternative expression for the variance of the mean of a systematic sample.

Theorem 4.3.1. *In systematic sampling,*

$$Var(\bar{y}_{sy}) = \frac{N-1}{N}S^2 - \frac{n-1}{n}S_w^2,$$

where

$$S^2 = \frac{\sum_{i=1}^{k}\sum_{j=1}^{n}(y_{ij} - \bar{y}_{\cdot\cdot})^2}{N-1}$$

= the mean sum of squares (MSS) for the whole population,

$$S_w^2 = \frac{\sum_{i=1}^{k}\sum_{j=1}^{n}(y_{ij} - \bar{y}_{i\cdot})^2}{n(k-1)}$$

= the mean sum of squares within systematic samples.

Proof. We have the following identity.

$$\sum_{i=1}^{k}\sum_{j=1}^{n}(y_{ij} - \bar{y}_{\cdot\cdot})^2 = \sum_{i=1}^{k}\sum_{j=1}^{n}(y_{ij} - \bar{y}_{i\cdot})^2 + n\sum_{i=1}^{k}(\bar{y}_{i\cdot} - \bar{y}_{\cdot\cdot})^2. \tag{4.3.1}$$

From (4.3.1), we have

$$(N-1)S^2 = k(n-1)S_w^2 + nk\,Var(\bar{y}_{sy}), \tag{4.3.2}$$

which gives

$$Var(\bar{y}_{sy}) = \frac{N-1}{N}S^2 - \frac{n-1}{n}S_w^2. \tag{4.3.3}$$

Thus, the theorem is proved.

The following theorem gives comparison of systematic sampling with SRSWOR.

Theorem 4.3.2. *A systematic sampling is more, equally or less efficient than SRSWOR accordingly as*

$$S_w^2 \gtreqless S^2.$$

Proof. For a SRSWOR of size n, we have

$$Var(\bar{y})_{SRS} = \frac{N-n}{Nn}S^2 = \frac{k-1}{N}S^2, \tag{4.3.4}$$

since $N = nk$. From (4.3.4) and (4.3.3), we obtain

$$Var(\bar{y})_{SRS} - Var(\bar{y}_{sy}) = \frac{n-1}{n}(S_w^2 - S^2), \tag{4.3.5}$$

from which the theorem follows.

From Theorem 4.3.2, we see that systematic sampling is more efficient than SRSWOR if the mean sum of squares within systematic samples is greater than the mean sum of squares for the whole population.

We now give another comparison of systematic sampling with SRSWOR in terms of intraclass correlation coefficient ρ.

Theorem 4.3.3. *In systematic sampling,*

$$Var(\bar{y}_{sy}) = \frac{N-1}{Nn}S^2[1 + (n-1)\rho],$$

where ρ is the intraclass correlation coefficient between pairs of units within systematic samples defined by

$$\rho = \frac{\dfrac{\sum_{i=1}^{k}\sum_{j\neq j'}^{n}(y_{ij} - \bar{y}_{..})(y_{ij'} - \bar{y}_{..})}{kn(n-1)}}{\dfrac{\sum_{i=1}^{k}\sum_{j=1}^{n}(y_{ij} - \bar{y}_{..})^2}{N}}.$$

Proof. Firstly, we note from the definition of ρ, that

$$(N-1)(n-1)\rho S^2 = \sum_{i=1}^{k}\sum_{j\neq j'}^{n}(y_{ij}-\bar{y}_{..})(y_{ij'}-\bar{y}_{..}). \tag{4.3.6}$$

Now, we have

$$n^2 k\, Var(\bar{y}_{sy}) = n^2\sum_{i=1}^{k}(\bar{y}_{i\cdot}-\bar{y}_{..})^2 = \sum_{i=1}^{k}\left[\sum_{j=1}^{n}(y_{ij}-\bar{y}_{..})\right]^2$$

$$= \sum_{i=1}^{k}\left[\sum_{j=1}^{n}(y_{ij}-\bar{y}_{..})^2 + \sum_{j\neq j'}^{n}(y_{ij}-\bar{y}_{..})(y_{ij'}-\bar{y}_{..})\right]$$

$$= (N-1)S^2 + (N-1)(n-1)S^2$$

$$= (N-1)S^2[1+(n-1)\rho], \tag{4.3.7}$$

from which the theorem follows.

Corollary 4.3.3.1.

$$-\frac{1}{(n-1)} \leq \rho \leq 1.$$

Theorem 4.3.4. *A systematic sampling is more, equally or less efficient than SRSWOR accordingly as*

$$\rho >=< -\frac{1}{(N-1)}.$$

Proof. We know that for SRSWOR of size n,

$$Var(\bar{y})_{SRS} = \frac{N-n}{Nn}S^2. \tag{4.3.8}$$

From Theorem 4.3.3 and (4.3.8), we obtain

$$Var(\bar{y})_{SRS} - Var(\bar{y}_{sy}) = -\frac{S^2}{Nn}(n-1)[1+\rho(N-1)]. \tag{4.3.9}$$

Theorem follows from (4.3.9).

Corollary 4.3.4.1. *The efficiency of systematic sampling relative to SRSWOR is given by*

$$E = \frac{N-n}{N-1}\cdot\frac{1}{[1+(n-1)\rho]}. \tag{4.3.10}$$

Theorem 4.3.5. *In systematic sampling,*

$$\rho = 1 - \frac{N}{N-1}\cdot\frac{S_w^2}{S^2},$$

where S_w^2 and S^2 are as defined in Theorem 4.3.1.

Proof. Proof follows by equating the results of Theorems 4.3.1 and 4.3.3.

4.4 Comparison with Stratified Sampling

In Section 4.1, we have seen the relation between systematic sampling and stratified sampling. In the equivalent stratified sampling, the population is divided into n strata each of size k and one unit is selected from each stratum. The columns in sample compositions of Table 4.1 form the strata. The jth stratum mean is $\bar{y}_{.j}$ defined by $\bar{y}_{.j} = \frac{1}{k}\sum_{i=1}^{k} y_{ij}$ and the jth mean sum of squares is S_j^2, defined by $S_j^2 = \frac{1}{k-1}\sum_{i=1}^{k}(y_{ij} - \bar{y}_{.j})^2$. Hence, noting $N = nk$, $N_j = k$, $W_j = \frac{N_j}{N} = \frac{1}{n}$, $n_j = 1$, and $S_j^2 = \frac{1}{k-1}\sum_{i=1}^{k}(y_{ij} - \bar{y}_{.j})^2$, we find that

$$Var(\bar{y}_{st}) = \sum_{j=1}^{n}\left(\frac{1}{n_j} - \frac{1}{N_j}\right) W_j^2 S_j^2$$

$$= \frac{N-n}{Nn} S_{wst}^2, \tag{4.4.1}$$

where

$$S_{wst}^2 = \frac{1}{n}\sum_{j=1}^{n} S_j^2$$

$$= \frac{1}{n(k-1)}\sum_{i=1}^{k}\sum_{j=1}^{n}(y_{ij} - \bar{y}_{.j})^2$$

$=$ mean sum of squares within strata.

Theorem 4.4.1. *In systematic sampling,*

$$Var(\bar{y}_{sy}) = \frac{N-n}{Nn} S_{wst}^2[1 + \rho_{wst}(n-1)],$$

where S_{wst}^2 is defined in (4.4.1) and ρ_{wst} is intraclass correlation coefficient between pairs of deviations of units which lie along the same row measured from their stratum means and is defined by

$$\rho_{wst} = \frac{\dfrac{\sum_{i=1}^{k}\sum_{j\neq j'}^{n}(y_{ij} - \bar{y}_{.j})(y_{ij'} - \bar{y}_{.j'})}{kn(n-1)}}{\dfrac{\sum_{i=1}^{k}\sum_{j=1}^{n}(y_{ij} - \bar{y}_{.j})^2}{nk}}.$$

Proof. From the definition of ρ_{wst} and S_{wst}^2 we obtain

$$(N-n)(n-1)S_{wst}^2 \cdot \rho_{wst} = \sum_{i=1}^{k}\sum_{j\neq j'}^{n}(y_{ij} - \bar{y}_{.j})(y_{ij'} - \bar{y}_{.j'}). \tag{4.4.2}$$

Now, we have

$$n^2k\,Var(\bar{y}_{sy}) = n^2\sum_{i=1}^{k}(\bar{y}_{i\cdot} - \bar{y}_{\cdot\cdot})^2$$

$$= \sum_{i=1}^{k}\left[\sum_{j=1}^{n}(y_{ij} - \bar{y}_{\cdot j})\right]^2$$

$$= \sum_{i=1}^{k}\sum_{j=1}^{n}(y_{ij} - \bar{y}_{\cdot j})^2$$

$$+ \sum_{i=1}^{k}\sum_{j\neq j'}^{n}(y_{ij} - \bar{y}_{\cdot j})(y_{ij'} - \bar{y}_{\cdot j'})$$

$$= (N-n)S^2_{wst} + (N-n)(n-1)S^2_{wst}\,\rho_{wst},$$

from which we obtain

$$Var(\bar{y}_{sy}) = \frac{N-n}{Nn}\cdot S^2_{wst}[1 + (n-1)\rho_{wst}]. \tag{4.4.3}$$

Theorem 4.4.2. *Systematic sampling is more, equally or less efficient than the corresponding equivalent stratified sampling accordingly as*

$$\rho_{wst} \gtreqless 0.$$

Proof. From (4.4.1) and (4.4.3), we obtain

$$Var(\bar{y}_{st}) - Var(\bar{y}_{sy}) = -\frac{N-n}{Nn}(n-1)\rho_{wst}\cdot S^2_{wst}, \tag{4.4.4}$$

from which the theorem follows.

Corollary 4.4.2.1. *The efficiency of systematic sampling relative to equivalent stratified sampling is given by*

$$E = \frac{1}{1 + (n-1)\rho_{wst}}. \tag{4.4.5}$$

4.5 Comparison of Systematic Sampling for Population with Linear Trends

We assume that the values of units in the population increase according to linear trend. We shall suppose that the values of the successive units increase by constant amount h. Note that y_{ij} defined in Section 4.2 represents the value of the $(i+(j-1)k)^{th}$ unit of the population. Hence, under the assumption of linear trend,

$$y_{ij} = \mu + h[i + (j-1)k], \quad i = 1, 2, \ldots, k; \quad j = 1, 2, \ldots, n.$$

The following lemma is very useful in deriving results of this section.

Lemma 4.5.1. *If $X_1, X_2, \ldots, X_M$ satisfy a linear trend, that is, they increase by a constant θ, X_i being equal to $a + \theta i$, $i = 1, 2, \ldots, M$; then*

$$\bar{X} = \frac{1}{M}\sum_{i=1}^{M} X_i = a + \frac{\theta(M+1)}{2} S_x^2 = \frac{\Sigma(X_i - \bar{X})^2}{M-1} = \frac{\theta^2 M(M+1)}{12}.$$

The proof is left to the reader as an exercise.

We now prove the following theorem giving the comparison of stratified sampling, systematic sampling and SRSWOR.

Theorem 4.5.1. *For populations with linear trend and $N = nk$,*

$$V_{st} : V_{sy} : V_{ran} :: \frac{1}{n} : 1 : \frac{nk+1}{k+1},$$

where V_{st}, V_{sy}, V_{ran} stand respectively for sampling variances of the sample unbised estimator of the population mean for stratified sampling, systematic sampling and SRSWOR.

Proof. We assume

$$y_{ij} = \mu + h[i + (j-1)k], \quad i = 1, 2, \ldots, k, \quad j = 1, 2, \ldots, n,$$

for the values of the units of the population.

(i) SRSWOR: Here, we know that

$$V_{ran} = Var(\bar{y}) = \frac{N-n}{Nn} S^2 = \frac{k-1}{N} S^2.$$

Now, using Lemma 4.5.1, we get

$$S^2 = \frac{h^2 N(N+1)}{12}.$$

Hence,

$$V_{ran} = \frac{(nk+1)(k-1)}{12} h^2. \tag{4.5.1}$$

(ii) Stratified Sampling: From (4.4.1), we have

$$V_{st} = Var(\bar{y}_{st}) = \frac{N-n}{Nn} S^2_{wst} = \frac{(k-1)}{nk} S^2_{wst}, \tag{4.5.2}$$

where

$$S^2_{wst} = \frac{1}{n} \sum_{j=1}^{n} S^2_j.$$

Now, if we consider the jth stratum, then noting that $y_{ij} = \mu + h[i + (j-1)k]$, we see that the value of jth unit increases by h. Hence, applying Lemma 4.5.1, we get

$$S^2_j = \frac{h^2 k(k+1)}{12},$$

$$S^2_{wst} = \frac{h^2 k(k+1)}{12} \tag{4.5.3}$$

and, from (4.5.2) and (4.5.3)

$$V_{st} = \frac{h^2(k^2-1)}{12n}. \tag{4.5.4}$$

(iii) Systematic Sampling: Here, we see that

$$\bar{y}_{i\cdot} = \mu + h\left[i + \frac{k(n-1)}{2}\right],$$

which shows that $\bar{y}_{i\cdot}$ increases by h. Hence, using Lemma 4.5.1 the mean sum of squares for means of systematic samples is given by

$$\frac{\sum_{i=1}^{k} (\bar{y}_{i\cdot} - \bar{y}_{\cdot\cdot})^2}{(k-1)} = \frac{h^2 k(k+1)}{12}. \tag{4.5.5}$$

Now, from (4.5.5), we find that

$$V_{sy} = \frac{1}{k} \sum_{i=1}^{k} (\bar{y}_{i\cdot} - \bar{y}_{\cdot\cdot})^2 = \frac{h^2(k^2-1)}{12}. \tag{4.5.6}$$

From (4.5.1), (4.5.4), and (4.5.6), the theorem follows.

Corollary 4.5.1. *If k is large, so that $\frac{1}{k}$ is negligible, then*

$$V_{st} : V_{sy} : V_{ran} :: \frac{1}{n} : 1 : n.$$

4.6 Estimation of Sampling Variance

In systematic sampling, it is not possible to estimate unbiasedly the variance. For, the systematic sample $\bar{y}_{sy}$ has the variance given by

$$Var(\bar{y}_{sy}) = \frac{1}{k}\sum_{i=1}^{k}(\bar{y}_{i\cdot} - \bar{y}_{\cdot\cdot})^2.$$

In systematic sampling, we have only one value $\bar{y}_{i\cdot}$ as the mean of a systematic sample, and from a sinngle value, variance cannot be estimated. However, in order to get some idea of the variance of a systematic mean, some biased estimators of the variance have been suggested. These are due to Cochran (1946) and Yates (1948). We give one of these below.

We consider the arrangement of systematic samples as in Table 4.1. Further, we consider $n = 2m$. Consider i^{th} systematic sample. Then, we assume that the $2m$ units in this sample are paired into m pairs, the tth pair being $(y_{i,2t-1}, y_{i,2t})$, $t = 1, 2, \ldots, m$. Let the mean of the tth pair be $\bar{y}_{i,t} = \frac{1}{2}(y_{i,2t-1} + y_{i,2t})$. We consider these two units as forming a SRSWOR from a population of $2k$ units. Now

$$\bar{y}_{i\cdot} = \frac{1}{m}\sum_{t=1}^{m}\bar{y}_{i,t}$$

and an unbiased estimator of variance of $\bar{y}_{i,t}$ is

$$\left(\frac{1}{2} - \frac{1}{2k}\right)\cdot\frac{(y_{i,2t-1} - y_{i,2t})^2}{2} = \frac{k-1}{4k}(y_{i,2t-1} - y_{i,2t})^2.$$

Hence, an estimator of variance of $\bar{y}_{i\cdot}$, i.e. of the mean of a systematic sample is given by

$$\begin{aligned} Est.Var(\bar{y}_{st}) &= \frac{1}{m^2}\cdot\frac{k-1}{4k}\sum_{t=1}^{m}(y_{i,2t-1} - y_{i,2t})^2 \\ &= \frac{k-1}{n^2k}\sum_{t=1}^{m}(y_{i,2t-1} - y_{i,2t})^2 \\ &= \frac{N-n}{Nn^2}\sum_{t=1}^{m}(y_{i,2t-1} - y_{i,2t})^2. \end{aligned}$$

4.7 Systematic Sampling when $N \neq nk$

Let $N = 11$ and $k = 3$. Then the different systematic samples are shown in Table 4.2.

Table 4.2 Systematic Samples with $k = 3$ and $N = 11$

Systematic Sample No.	Systematic Sample	Prob.	Sample Mean
1	Y_1, Y_4, Y_7, Y_{10}	$\frac{1}{3}$	$\frac{1}{4}(Y_1 + Y_4 + Y_7 + Y_{10})$
2	Y_2, Y_5, Y_8, Y_{11}	$\frac{1}{3}$	$\frac{1}{4}(Y_2 + Y_5 + Y_8 + Y_{11})$
3	Y_3, Y_6, Y_9	$\frac{1}{3}$	$\frac{1}{3}(Y_3 + Y_6 + Y_9)$

From the above example, we note two things:

(i) Sample size is not same.

(ii) Sample mean is not an unbiased estimator of the population mean $\bar{Y}$, since

Expected Value of Sample Mean

$$= \frac{1}{3}\left[\frac{1}{4}(Y_1 + Y_4 + Y_7 + Y_{10}) + \frac{1}{4}(Y_2 + Y_5 + Y_8 + Y_{11}) + \frac{1}{3}(Y_3 + Y_6 + Y_9)\right]$$

$$\neq \frac{1}{11}\sum_{i=1}^{11} Y_i = \bar{Y}\ .$$

However, it is possible to get an unbiased estimator of the population mean. This is done as follows in the general case.

Suppose N is not a multiple of n. Then the number of units selected systematically with the sampling interval k equal to the nearest integer to $\frac{N}{n}$ may not be necessarily equal to n. Let $N = nq + r$. Then, we take the sampling interval k to be q or $q + 1$ accordingly as $r \leq \frac{n}{2}$ or $r > \frac{n}{2}$. If k is q' ($= q$ or $q + 1$), then the number of units expected in the sample would be equal to $\left[\frac{N}{q'}\right]$ or $\left[\frac{N}{q'}\right] + 1$ with probabilities $\left[\frac{N}{q'}\right] + 1 - \left(\frac{N}{q'}\right)$ and $\left(\frac{N}{q'}\right) - \left[\frac{N}{q'}\right]$ respectively, where $\left[\frac{M}{g}\right]$ notation denotes the largest integer contained in $\frac{M}{g}$. If $q' = q$, then we get $n' = n + \left[\frac{r}{q}\right]$ and $n' = n + \left[\frac{r}{q}\right] + 1$ with probabilities $\left(\frac{r}{q}\right) + 1 - \left[\frac{r}{q}\right]$ and $\left(\frac{r}{q}\right) - \left[\frac{r}{q}\right]$ respectively. Similarly, if $q' = q + 1$, then we obtain $n' = n - \left(\frac{n-r}{q+1}\right)$ and $n' = n - \left[\left(\frac{n-r}{q+1}\right) + 1\right]$ with probabilities $\left[\frac{(n-r)}{(q+1)}\right] + 1 - \left(\frac{n-r}{q+1}\right)$ and $\left(\frac{n-r}{q+1}\right) - \left[\frac{(n-r)}{(q+1)}\right]$ respectively.

Example 4.1. Let $N = 17$ and $n = 5$. Then $q = 3$ and $r = 2$. Since $r < \frac{n}{2}$, we take $k = q = 3$. Then, the sample sizes would be $n' = n + \left[\frac{r}{q}\right] = 5$, and

Table 4.3 Systematic Samples from Population with $N = 17$ and $k = 3$

Systematic Sample No.	Systematic Sample	Probability
1	$Y_1, Y_4, Y_7, Y_{10}, Y_{13}, Y_{16}$	$\frac{1}{3}$
2	$Y_2, Y_5, Y_8, Y_{11}, Y_{14}, Y_{17}$	$\frac{1}{3}$
3	$Y_3, Y_6, Y_9, Y_{12}, Y_{15}$	$\frac{1}{3}$

$n' = n + \left[\frac{r}{q}\right] + 1 = 6$ with respective probabilities $\left[\frac{r}{q}\right] + 1 - \left(\frac{r}{q}\right) = \frac{1}{3}$ and $\left(\frac{r}{q}\right) - \left[\frac{r}{q}\right] = \frac{2}{3}$. This can be verified from Table 4.3.

We now prove the following theorem which shows how to obtain an unbiased estimator of the population mean when $N \neq nk$.

Theorem 4.7.1. *In systematic sampling with sampling interval k from a population with size $N \neq nk$, an unbiased estimator of the population mean $\bar{Y}$ is given by*

$$\hat{\bar{Y}} = \frac{k}{N}\left(\sum^{n'} y\right)_i$$

where i stands for the ith systematic sample $i = 1, 2, \ldots, k$ and n' denotes the size of the ith systematic sample.

Proof. Each systematic sample has probability $\frac{1}{k}$. Hence,

$$E(\hat{\bar{Y}}) = \sum_{i=1}^{k} \frac{1}{k} \cdot \frac{k}{N}\left(\sum^{n'} y\right)_i$$

$$= \frac{1}{N}\sum_{i=1}^{k}\left(\sum^{n'} y\right)_i.$$

Now, each unit occurs in only one of the k possible systematic samples. Hence

$$\sum_{i=1}^{k}\left(\sum^{n'} y\right)_i = \sum_{i=1}^{N} Y_i\,,$$

which on substitution in (4.7.1) proves the theorem.

4.8 Circular Systematic Sampling

When $N \neq nk$, we have seen in Section 4.7, that the systematic samples are not of the same size and the sample mean is not an unbiased estimator of the population mean. To overcome these disadvantages of systematic sampling when $N \neq nk$, circular systematic sampling is proposed. Circular systematic sampling consists in selecting a random number from 1 to N and selecting the unit corresponding to this random number and thereafter every k^{th} unit in a cyclical manner till a sample of n units is obtained, k being the nearest integer to $\frac{N}{n}$. In other words, if i is a number selected at random from 1 to N, then the circular systematic sample consists of units with serial numbers

$$\left.\begin{array}{rl} i+jk, & \text{if } i = jk \leq N \\ i+jk-N, & \text{if } i+jk > N \end{array}\right\} \qquad j = 0, 1, 2, ..., (n-1)$$

This sampling scheme ensures equal probability of inclusion in the sample for every unit and was suggested by Lahiri (1952).

Example 4.2. Let $N = 14$ and $n = 5$. Then, k = nearest integer to $\frac{14}{5} = 3$. Let the first number selected at random from 1 to 14 be 7. Then, the circular systematic sample consists of units with serial numbers

$$7, 10, 13, \qquad 16 - 14 = 2, \qquad 19 - 14 = 5.$$

This procedure is illustrated diagrammatically in Figure 4.1.

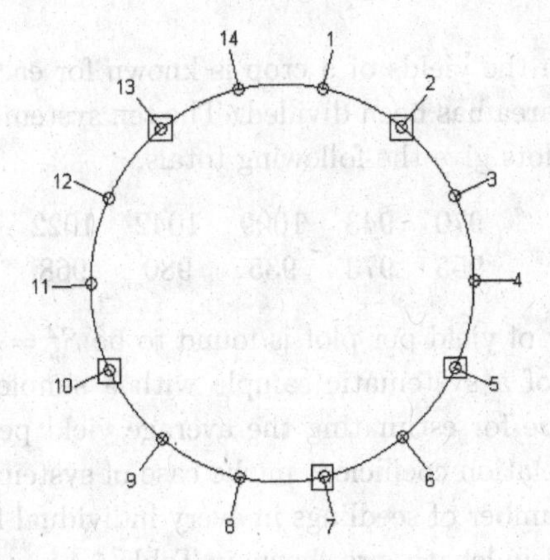

Fig. 4.1 Diagrammatic representation of circular systematic sampling.

Theorem 4.8.1. *In circular systematic sampling, the sample mean is an unbiased estimator of the population mean.*

Proof. If i is the number selected at random, then the circular systematic sample mean is

$$\bar{y} = \frac{1}{n}\left(\sum^{n} y\right)_i ,$$

where $\left(\sum^{n} y\right)_i$ denotes the total of y values in the ith circular systematic sample, $i = 1, 2, \ldots, N$. We note here that in circular systematic sampling, there are N circular systematic samples, each having probability $\frac{1}{N}$ of its selection. Hence,

$$E(\bar{y}) = \sum_{i=1}^{N} \frac{1}{n}\left(\sum^{n} y\right)_i \times \frac{1}{N} = \frac{1}{Nn}\sum_{i=1}^{N}\left(\sum^{n} y\right)_i . \tag{4.8.1}$$

Clearly, each unit of the population occurs in n of the N possible circular systematic sample means. Hence,

$$\sum_{i=1}^{N}\left(\sum^{n} y\right)_i = n\sum_{i=1}^{N} Y_i,$$

which on substitution in (4.8.1) proves the theorem.

EXERCISES

4.1. Information on the yields of a crop is known for each of the 80 plots into which an area has been divided. The ten systematic samples each containing 8 plots give the following totals.

970	943	1009	1042	1022
955	973	935	980	968

The variability of yield per plot is found to be $S_y^2 = 107.57$. Compare the efficiency of a systematic sample with a simple random sample of the same size for estimating the average yield per plot. Find the intraclass correlation coefficient in the case of systematic sampling.

4.2. Data on the number of seedlings in every individual foot of sown bed, which is 60 feet in length, are shown in Table 4.4 in a rectangular form for convenience.

Table 4.4

1	2	3	4	5	6	7	8	9	10	11	12	13
M	M	M	M	M	M	M	M	M	M	M	M	M
F	F	F	F	F	F	F	F	F	F	F	F	F
m	f		m	f	m	m	f		f	m		
f			f	m		f						

(a) Find the relative standard error of the estimator of the total number of seedlings based on a systematic sample consisting of every 10th foot of the sown bed.

(b) Determine the relative efficiency of systematic sampling compared with SRSWOR when the sample size is 6 one-foot bed lengths.

(c) Determine the relative efficiency of systematic sampling compared with stratified sampling when one one-foot bed length is selected from each column, columns being treated as strata.

4.3. Suppose it is assumed that a finite population of N units is drawn from the super-populations with the following models:

(i) $E(Y_i) = \mu$, $Var(Y_i) = \sigma_i^2$, $Cov(Y_i, Y_j) = 0$, $i \neq j$.

(ii) $E(Y_i) = \alpha + i\beta$, $Var(Y_i) = \sigma^2$, $Cov(Y_i, Y_j) = 0$, $i \neq j$.

For each of the above two cases, derive the expressions for the expected variances of the sample mean $\bar{y}$ based on samples of size n selected (a) systematically and (b) with SRSWOR, when N is a multiple of n, and compare them.

4.4. Suppose in a population with linear trend, the observations are assumed to be independentlly distributed with the same variance. Show that the variance of the estimator based on a systematic sample is inflated on an average by applying the end corrections to the extent of

$$\frac{k^2 - 1}{6(n-1)k^2},$$

where k is the sampling interval.

4.5. Let $V_{sy}(1)$ denote the variance of the estimator based on a systematic sample of size n, (1) indicating one random start; $V_{st}(1)$ denote the variance of the estimator based on a stratified random sample where one unit is independently drawn from each of the n strata $\{y_1, y_2, \ldots, y_k\}, \{y_{k+1}, \ldots, y_{2k}\}, \ldots$, and $V_R(n)$ denote the variance of the estimator based on a SRSWOR of size n. In order to overcome the difficulty of estimating the variance of a systematic sampling, it has been recommended to select a systematic sample of size ns with s random starts. Let $V_{sy}(s)$ denote the variance of the estimator based on a systematic sample of size ns with s random starts, $V_{st}(s)$ denote the variance of the estimator based on a stratified random sample when s units are independently drawn from each stratum and $V_R(ns)$ denote the variance of the estimator based on a SRSWOR of size ns. Show that the relative magnitudes of the three variances $V_{sy}(1)$, $V_{st}(1)$ and $V_R(n)$ are the same as for $V_{sy}(s)$, $V_{st}(s)$, and $V_R(ns)$ respectively.

4.6. A population contains $N = nk$ units where k is odd. Let $\bar{y}_i$ denote the mean of the systematic sample based on the random start i taken between 1 and k. The centered systematic sample estimator will be obtained by taking the mean of the central units numbered $[\frac{k+1}{2}, k+\frac{k+1}{2}, \ldots]$ from each of the n strata formed when a random-start systematic sample is selected. If the population is monotone increasing, the centered systematic sample mean $\bar{y}_c$ will be the median of the k random-start systematic sample means $\bar{y}_1 < \bar{y}_2 < \ldots < \bar{y}_k$. The mean square error of $\bar{y}_c$ will be $(\bar{y}_c - \bar{Y})^2$ while the variance of $\bar{y}_i$ would be $\frac{1}{k}\sum_{i=1}^{k}(\bar{y}_i - \bar{Y})^2$. Use the result (mean-median) $<$ variance, to prove that centered systematic sampling is more efficient than random-start systematic sampling in the case of monotone populations.

4.7. In a directory of 13 houses on a street, the persons are listed as follows: M = male adult, F = female adult, m = male child, f = female child. Compare the variances given by a systematic sample of one in five persons and a 20% SRSWOR for estimating (i) the proportion of males, (ii) the proportion of children, (iii) the proportion of persons living in professional households (1, 2, 4, 6, 13 are described as professional). For systematic sample, number down each column, then go to the top of the next column.

4.8. In a population with a linear trend, show that a systematic sample is less precise than a stratified sample with strata of size $2k$ and two units per stratum if $n > \frac{4k+2}{k+1}$.

Chapter 5

RATIO METHOD OF ESTIMATION

5.1 Ratio Method of Estimation

Sometimes in sample surveys, along with the information on the study variable y, information on auxiliary variate x, correlated with y, is also collected. This information on auxiliary variate x, may be utilized to obtain a more efficient estimator of the population mean. Ratio method of estimation is an attempt in this direction.

Consider a population of N units. Let $\bar{Y}$ and $\bar{X}$ be the means of y and x characters respectively, and $R = \frac{\bar{Y}}{\bar{X}}$ be the ratio of the population means $\bar{Y}$ and $\bar{X}$. We assume that a simple random sample without replacement of size n has been drawn from the above population and values of y and x characters are measured for each of the n units in the sample. Let $\bar{y}$ and $\bar{x}$ be the sample means of y and x characters respectively. Further, we assume that $\bar{X}$ is known.

We now define an estimator of R as

$$\hat{R} = \frac{\bar{y}}{\bar{x}}, \tag{5.1.1}$$

and a ratio estimator of $\bar{Y}$ as

$$\bar{y}_R = \hat{R} \cdot \bar{X} = \frac{\bar{y}}{\bar{x}}\bar{X}. \tag{5.1.2}$$

This method of estimation may be advantageously used when (i) x represents the same character as y, but measured at some previous date when a complete count of the population was made and (ii) the character x is cheaply, quickly and easily measurable.

5.2 Bias of Ratio Estimator

Unlike other estimators discussed in previous chapters, we shall see here that the ratio estimator $\bar{y}_R$ is generally a biased estimator. It is to be noted that a biased esstimator may be preferred to an unbiased estimator if the mean square error of the former is less than the variance of the latter.

We shall derive here an expression for the bias of the ratio estimator $\bar{y}_R$. We use the following notations

$$e_1 = \frac{\bar{x} - \bar{X}}{\bar{X}}, \qquad e_2 = \frac{\bar{y} - \bar{Y}}{\bar{Y}}. \tag{5.2.1}$$

From the definitions of e_1 and e_2, we obtain

$$\begin{aligned} E(e_1) = 0, \quad Var(e_1) = E(e_1^2) &= \frac{Var(\bar{x})}{\bar{X}^2} \\ E(e_2) = 0, \quad Var(e_2) = E(e_2^2) &= \frac{Var(\bar{y})}{\bar{Y}^2} \\ Cov(e_1, e_2) = E(e_1, e_2) &= \frac{Cov(\bar{y}, \bar{x})}{\bar{X}\bar{Y}}. \end{aligned} \tag{5.2.2}$$

We make the following two assumptions:

(i) $|e_1| < 1$, i.e. $0 < \bar{x} < 2\bar{X}$, $\bar{X}$ is assumed to be positive.
(ii) n is large enough so that terms in (e_1, e_2) of degree greater than 2 are negligible.

We now prove the following theorem.

Theorem 5.2.1. *Approximate bias in the ratio estimator $\bar{y}_R$ of $\bar{Y}$ is given by*

$$Bias(\bar{y}_R) = \frac{1}{\bar{X}}[R \cdot Var(\bar{x}) - Cov(\bar{y}, \bar{x})].$$

Proof. We have

$$\begin{aligned} \bar{y}_R = \frac{\bar{y}}{\bar{x}}\bar{X} &= \bar{Y}(1 + e_2)(1 + e_1)^{-1} \\ &= \bar{Y}(1 + e_2)(1 - e_1 + e_1^2 \ldots) \\ &= \bar{Y}[1 - e_1 + e_1^2 + e_2 - e_1 e_2], \end{aligned} \tag{5.2.3}$$

since other terms are negligible. Taking expectation of (5.2.3) and using (5.2.2), we obtain

$$E(\bar{y}_R) = \bar{Y}\left[1 + \frac{Var(\bar{x})}{\bar{X}^2} - \frac{Cov(\bar{y}, \bar{x})}{\bar{X}\bar{Y}}\right], \tag{5.2.4}$$

from which the theorem follows.

Corollary 5.2.1.

$$Bias(\bar{y}_R) = \frac{1-f}{n\bar{x}}[R \cdot S_x^2 - S_{xy}],$$

where $f = \frac{n}{N}$, $S_x^2 = \sum_{i=1}^{N} \frac{(x_i - \bar{X})^2}{(N-1)}$, $S_{xy} = \sum_{i=1}^{N} \frac{(x_i - \bar{X})(y_i - \bar{Y})}{(N-1)}$.

Corollary 5.2.2. *Approximate expression of bias in ratio estimator* $\bar{y}_R$ *becomes zero if the regression of* y *on* x *is a straight line passing through the origin.*

For, if the line of regression of y on x is a straight line passing through the origin, then we have

$$y = \beta x,$$

from which one obtains $\bar{Y} = \beta\bar{X}$. But

$$\beta = \frac{Cov(y,x)}{Var(x)} = \frac{S_{xy}}{S_x^2}.$$

Hence, $\bar{Y} = \frac{S_{xy}}{S_x^2}\bar{X}$, i.e. $R \cdot S_x^2 = S_{xy}$. Therefore, the expression for bias derived in Theorem 5.2.1 becomes zero, according to Corollary 5.2.1.

Corollary 5.2.3.

$$Bias(\hat{R}) = \frac{1}{\bar{X}^2}[R \cdot Var(\bar{x}) - Cov(\bar{y}, \bar{x})].$$

This follows from the fact that $\hat{R} = \frac{1}{\bar{X}}\bar{y}_R$.

Corollary 5.2.4. *Bias in the ratio estimator* $\bar{y}_R$ *is estimated by*

$$Est.Bias(\bar{y}_R) = \frac{1-f}{n\bar{X}}[\hat{R} \cdot s_x^2 - s_{xy}].$$

Without making any assumption, we give an exact expression of bias of the ratio estimator $\bar{y}_R$ in the following theorem.

Theorem 5.2.2.

$$Bias(\bar{y}_R) = -Cov(\hat{R}, \bar{x}).$$

Proof. We have

$$\begin{aligned} -Cov(\hat{R},\bar{x}) &= -E(\hat{R}\bar{x}) + E(\hat{R})E(\bar{x}) \\ &= -E(\bar{y}_R) + E(\hat{R}\bar{X}) \\ &= E(\bar{y}_R) - \bar{Y} \\ &= Bias(\bar{y}_R), \end{aligned}$$

which proves the theorem.

Remark 1. From Corollary 5.2.1, we note that bias decreases as the sample size n increases.

Remark 2. An alternative expression for $Bias(\bar{y}_R)$ can be derived as follows.

$$Bias(\bar{y}_R) = -Cov(\hat{R},\bar{x}) = -\rho(\hat{R},\bar{x})\sigma_{\bar{x}} \cdot \sigma_{\hat{R}}, \tag{5.2.5}$$

where $\rho(\hat{R},\bar{x})$ is the coefficient of correlation between $\hat{R}$ and $\bar{x}$; and $\sigma_{\bar{x}}$ and $\sigma_{\hat{R}}$ are the standard deviations of $\bar{x}$ and $\hat{R}$ respectively. Hence, relative bias of $\bar{y}_R$ is given by

$$\begin{aligned} \frac{Bias(\bar{y}_R)}{\bar{Y}} &= -\rho(\hat{R},\bar{x}) \cdot \frac{\sigma_{\bar{x}}}{\bar{X}} \cdot \frac{\sigma_{\hat{R}}}{R} \\ &= -\rho(\hat{R},\bar{x})C(\bar{x})C(\hat{R}), \end{aligned} \tag{5.2.6}$$

where $C(\bar{x})$ and $C(\hat{R})$ are the relative standard errors of $\bar{x}$ and $\hat{R}$ respectively.

From (5.2.6) one obtains

$$\left|\frac{Bias(\bar{y}_R)}{\bar{Y}}\right| = \rho(\hat{R},\bar{x}) \cdot C(\bar{x})C(\hat{R}). \tag{5.2.7}$$

When n is large, $C(\bar{x})$ and $C(\hat{R})$ are small, and hence relative bias in $\bar{y}_R$ would be small. Further, an upper bound for the magnitude of the relative bias is given by

$$\left|\frac{Bias(\bar{y}_R)}{\bar{Y}}\right| \leq C(\bar{x}) \cdot C(\hat{R}). \tag{5.2.8}$$

5.3 Approximate Variance of Ratio Estimator

Since ratio estimator is a biased estimator, we shall work out its approximate variance under the assumptions made in Section 5.2. For the sake of convenience, we call the approximate variance of ratio estimator simply variance with the understanding that it is only approximate.

Theorem 5.3.1. *The approximate variance of the ratio estimator* $\bar{y}_R$ *is given by*

$$Var(\bar{y}_R) = \frac{1-f}{n}[S_y^2 + R^2 S_x^2 - 2RS_{xy}],$$

where $f = \frac{n}{N}$, S_y^2, S_x^2 *are mean sum of squares for the characters* y *and* x *respectively and* S_{xy} *is defined by*

$$S_{xy} = \sum_{i=1}^{N} \frac{(x_i - \bar{X})(y_i - \bar{Y})}{(N-1)}.$$

Proof. We have

$$\begin{aligned} \bar{y}_R &= \frac{\bar{y}}{\bar{x}}\bar{X} = \bar{Y}(1+e_2)(1+e_1)^{-1} \\ &= \bar{Y}[(1+e_2)(1-e_1+e_1^2\ldots)] \\ &= \bar{Y}[1-e_1+e_1^2+e_2-e_2e_1], \end{aligned} \tag{5.3.1}$$

since terms in (e_1, e_2) of degree greater than 2 are assumed to be negligible. Then, approximate variance of $\bar{y}_R$ is given by

$$\begin{aligned} Var(\bar{y}_R) = E(\bar{y}_R - \bar{Y})^2 &= E[\bar{Y}^2(-e_1+e_1^2+e_2-e_2e_1)^2] \\ &= \bar{Y}^2 E(e_2^2 + e_1^2 - 2e_1e_2), \\ &= Var(\bar{y}) + R^2 Var(\bar{x}) - 2R\,Cov(\bar{x}, \bar{y}) \end{aligned} \tag{5.3.2}$$

since terms in (e_1, e_2) of degree greater than 2 are negligible. Then, using (5.2.2), we have from (5.3.2)

$$Var(\bar{y}_R) = \frac{1-f}{n}[S_y^2 + R^2 S_x^2 - 2R\,S_{xy}],$$

which proves the theorem.

Remark. Alternative expressions for $Var(\bar{y}_R)$ are as follows:

$$Var(\bar{y}_R) = \frac{1-f}{n}[S_y^2 + R^2 S_x^2 - 2R\rho S_x S_y]. \tag{5.3.4}$$

$$Var(\bar{y}_R) = \frac{1-f}{n} \cdot \bar{Y}^2[C_y^2 + C_x^2 - 2\rho C_x C_y], \tag{5.3.4}$$

where ρ is the coefficient of correlation between the characters y and x and C_y and C_x are respectively their coefficients of variation defined by $C_y = \frac{S_y}{\bar{Y}}$, $C_x = \frac{S_x}{\bar{X}}$.

Theorem 5.3.2. *An estimator of the variance of ratio estimator of the population mean is provided by*

$$Est.Var(\bar{y}_R) = \frac{1-f}{n(n-1)} \sum_{i=1}^{n} (y_i - \hat{R}x_i)^2.$$

Proof. Since, s_y^2, s_x^2 and s_{xy} are unbiased estimators of S_y^2, S_x^2 and S_{xy} respectively, and $\hat{R}$ an estimator of R, we obtain an estimator of $Var(\bar{y}_R)$ as

$$Est.Var(\bar{y}_R) = \frac{1-f}{n}[s_y^2 + \hat{R}^2 s_x^2 - 2\hat{R} s_x s_y]$$

$$= \frac{1-f}{n(n-1)} \sum_{i=1}^{n} (y_i - \hat{R}x_i)^2,$$

which proves the theorem.

We may note that (5.3.5) is not an unbiased estimator of $Var(\bar{y}_R)$.

5.4 Comparison with SRSWOR

Here, we shall compare the efficiency of ratio estimator $\bar{y}_R$ with that of the mean of a SRSWOR.

Theorem 5.4.1. *Ratio estimator is more efficient than the mean of a simple random sample without replacement if*

$$\rho > \frac{1}{2} \cdot \frac{C_x}{C_y}, \quad \text{if } R > 0$$

and

$$\rho < -\frac{1}{2} \cdot \frac{C_x}{C_y}, \quad \text{if } R < 0.$$

Proof. The variance of the mean of a SRSWOR is given by

$$Var(\bar{y})_{SRS} = \frac{1-f}{n} \cdot S_y^2. \tag{5.4.1}$$

The variance of the ratio estimator $\bar{y}_R$ is given by

$$Var(\bar{y}_R) = \frac{1-f}{n}[S_y^2 + R^2 S_x^2 - 2R\, S_x S_y]. \tag{5.4.2}$$

From (5.4.1) and (5.4.2), one obtains

$$Var(\bar{y})_{SRS} - Var(\bar{y}_R) = \frac{1-f}{n}[2\rho S_y - RS_x]RS_x. \tag{5.4.3}$$

Then, when $R > 0$, we deduce from (5.4.3), that ratio estimator is more efficient than $\bar{y}$ if

$$2\rho S_y - RS_x > 0 \quad \text{i.e. } \rho > \frac{1}{2}R\frac{S_x}{S_y}; \text{ i.e. } \rho > \frac{1}{2} \cdot \frac{C_x}{C_y}. \tag{5.4.4}$$

Further, when R is negative, then ratio estimator is more efficient than $\bar{y}$ if

$$2\rho S_y - RS_x < 0 \quad \text{i.e. } \rho < \frac{1}{2}R\frac{S_x}{S_y}; \text{ i.e. } \rho < -\frac{1}{2} \cdot \frac{C_x}{C_y}. \tag{5.4.5}$$

Thus, the theorem is established.

Remark. Usually R is > 0. So, ratio estimator is more efficient than $\bar{y}$ if $\rho > \frac{1}{2} \cdot \frac{C_x}{C_y}$. Further, when x is the same character as y but measured at some other time, then C_x and C_y are nearly equal. In this situation, ratio estimator is more efficient than $\bar{y}$ if $\rho > \frac{1}{2}$.

5.5 Estimation of a Ratio

In many surveys, one is interested in the ratio of population totals or means of two characters. For instance, in socio-economic surveys, one may be interested in ratios such as per capita income, proportion of unemployed persons, sex-ratio, birth-rates, death rates, etc. Similarly, estimation of the yield rate in a crop survey is important.

Let $R = \frac{\bar{Y}}{\bar{X}}$ be the ratio of population means of two characters y and x. Then, we can estimate R by the following three estimators.

$$\hat{R} = \frac{\bar{y}}{\bar{x}}, \quad \hat{R}_1 = \frac{\bar{y}}{\bar{X}}, \quad \hat{R}_2 = \frac{1}{n}\sum_{i=1}^{n}\frac{y_i}{x_i}.$$

We may note that $\hat{R}$ and $\hat{R}_2$ are biased estimators of R, while $\hat{R}_1$ is an unbiased estimator of R. We assume here SRSWOR. Now from Corollary 5.2.3, we have

$$\begin{aligned} Bias(\hat{R}) &= \frac{1}{\bar{X}^2}[R \cdot Var(\bar{x}) - Cov(\bar{y}, \bar{x})] \\ &= \frac{1-f}{n\bar{X}^2}[R \cdot S_x^2 - S_{xy}], \end{aligned} \tag{5.5.1}$$

which shows that the bias of $\hat{R}$ decreases as n increases.

Consider now $\hat{R}_2$. Bias of $\hat{R}_2$ is obtained as

$$Bias(\hat{R}_2) = E(\hat{R}_2) - R$$

$$= \frac{1}{n} \cdot \frac{\binom{N-1}{n-1} \sum_{i=1}^{N} \frac{y_i}{x_i}}{\binom{N}{n}} - R$$

$$= \frac{1}{N} \sum_{i=1}^{N} \frac{y_i}{x_i} - R, \quad (5.5.2)$$

which clearly does not depend upon n. Hence $Bias(\hat{R}_2)$ does not decrease as n increases. Hence, $\hat{R}_2$ is not a good choice for estimation of R.

We now consider the variances of $\hat{R}$ and $\hat{R}_1$. For $\hat{R}$, it will be its approximate variance. We note that $\hat{R} = \frac{\bar{y}_R}{\bar{X}}$, hence the variance of $\hat{R}$ will be obtained by dividing $Var(\bar{y}_R)$ by $\bar{X}^2$. Therefore, from (5.3.4), we obtain the variance of $\hat{R}$ as

$$Var(\hat{R}) = \frac{1-f}{n\bar{X}^2}[S_y^2 + R^2 S_x^2 - 2R\rho S_x S_y]. \quad (5.5.3)$$

Further, variance of $\hat{R}_1 = \frac{\bar{y}}{\bar{X}}$ is given by

$$Var(\hat{R}_1) = \frac{1-f}{n\bar{X}^2} \cdot S_y^2. \quad (5.5.4)$$

Thus, the comparison of variances of $\hat{R}$ and $\hat{R}_1$ will follow from that of the variances of $\bar{y}_R$ and $\bar{y}$. Hence, from Theorem 5.4.1, we deduce that $\hat{R}$ is more efficient than $\hat{R}_1$ if

$$\rho > \frac{1}{2} \cdot \frac{C_x}{C_y}, \quad \text{if } R > 0$$

and

$$\rho < -\frac{1}{2} \cdot \frac{C_x}{C_y}, \quad \text{if } R < 0.$$

We can estimate the variance of R by

$$Est.Var(\hat{R}) = \frac{1-f}{n(n-1)\bar{X}^2} \sum_{i=1}^{n} (y_i - \hat{R}x_i).$$

5.6 Conditions under which Ratio Estimator is the Best

Under certain conditions ratio estimator is the best linear unbiased estimator. The conditions are given in the following theorem. We consider our sample to be drawn from an infinite population.

Theorem 5.6.1. *The ratio estimator $\hat{R}$ is the best linear unbiased estimator of R if the following two conditions hold:*

(i) *For fixed x, $E(y) = \beta x$, i.e. the line of regression of y on x is a straight line passing through the origin.*

(ii) *For fixed x, $Var(x) \propto x$, i.e. $Var(x) = \lambda x$, where λ is the constant of proportionalilty.*

Proof. Let $\mathbf{y}' = (y_1, y_2, \ldots, y_n)$ and $\mathbf{x}' = (x_1, x_2, \ldots, x_n)$. Hence, for fixed $\mathbf{x}$,

$$E(\mathbf{y}) = \beta \mathbf{x}$$

$$Var(\mathbf{y}) = V = \lambda\, diag(x_1, x_2, \ldots, x_n).$$

Therefore, the best linear unbiased estimator of β is obtained by minimizing

$$(\mathbf{y} - \beta\mathbf{x})' V^{-1} (\mathbf{y} - x) \quad \text{i.e.} \sum_{i=1}^{n} \frac{(y_i - \beta x_i)^2}{\lambda x_i}. \tag{5.6.1}$$

Equating the derivative of (5.6.1) w.r.t. β to zero, we get

$$\sum_{i=1}^{n} (y_i - \hat{\beta} x_i) = 0,$$

which gives

$$\hat{\beta} = \frac{\bar{y}}{\bar{x}} = \hat{R}.$$

Thus, $\hat{R}$ is the best linear estimator of R. Consequently $\hat{R}\bar{X} = \bar{y}_R$ is the best linear unbiased estimator of $\bar{Y}$.

5.7 Ratio Estimation in Stratified Sampling

We now consider the ratio method of estimation in stratified sampling. We consider SRSWOR of size n_i from the ith stratum, $i = 1, 2, \ldots, k$. There are two types of ratio estimators which can be defined in stratified sampling. They are as follows:

(i) Combined Ratio Estimator: When $\bar{X}$ is known, we define the combined ratio estimator $\hat{\bar{Y}}_{CR}$ as

$$\hat{\bar{Y}}_{CR} = \frac{\bar{y}_{st}}{\bar{x}_{st}} \bar{X}, \tag{5.7.1}$$

where

$$\bar{y}_{st} = \sum_{i=1}^{k} w_i \bar{y}_i, \quad \bar{x}_{st} = \sum_{i=1}^{k} w_i \bar{x}_i,$$

and

$$w_i = \frac{N_i}{N}.$$

(ii) Separate Ratio Estimator: When stratum means $\bar{X}_i$, $i = 1, 2, \ldots, k$ are known, we define the separate ratio estimator $\hat{\bar{Y}}_{CR}$ as

$$\hat{\bar{Y}}_{SR} = \sum_{i=1}^{k} w_i \hat{\bar{Y}}_{iR}, \tag{5.7.2}$$

where

$$\hat{\bar{Y}}_{iR} = \frac{\bar{y}_i}{\bar{x}_i}\bar{X}_i = \text{ratio estimator of the stratum mean } \bar{Y}_i, \quad i = 1, 2, \ldots, k.$$

Theorem 5.7.1. *The bias of the combined ratio estimator $\bar{Y}_{CR}$ is given by*

$$Bias(\hat{\bar{Y}}_{CR}) = \frac{1}{\bar{X}} \sum_{i=1}^{k} w_i^2 \frac{(1-f_i)}{n_i}(RS_{ix}^2 - S_{ixy}).$$

Proof. We define e_1 and e_2 as

$$e_1 = \frac{\bar{x}_{st} - \bar{X}}{\bar{X}} \quad \text{and} \quad e_2 = \frac{\bar{y}_{st} - \bar{Y}}{\bar{Y}}.$$

Then, one obtains

$$\begin{aligned} E(e_1) = 0, \quad Var(e_1) &= \tfrac{1}{\bar{X}^2} Var(\bar{x}_{st}) \\ E(e_2) = 0, \quad Var(e_2) &= \tfrac{1}{\bar{Y}^2} Var(\bar{y}_{st}) \\ E(e_1\, e_2) &= Cov(\bar{x}_{st}, \bar{y}_{st}). \end{aligned} \tag{5.7.3}$$

Now, we have

$$\begin{aligned} \hat{\bar{Y}}_{CR} &= \frac{\bar{y}_{st}}{\bar{x}_{st}} \bar{X} \\ &= \bar{Y}(1+e_2)(1+e_1)^{-1} \\ &= \bar{Y}[1 - e_1 + e_1^2 + e_2 - e_2 e_1], \end{aligned} \tag{5.7.4}$$

assuming terms in (e_1, e_2) of degree greater than two are negligible.

Taking expectation of (5.7.4) and using (5.7.3), we obtain

$$E(\hat{\bar{Y}}_{CR}) = \bar{Y}\left[1 + \frac{1}{\bar{X}^2} Var(\bar{x}_{st}) - \frac{1}{\bar{X}\bar{Y}} Cov(\bar{x}_{st}, \bar{y}_{st})\right],$$

from which we obtain bias in $\hat{\bar{Y}}_{CR}$ as

$$Bias(\hat{\bar{Y}}_{CR}) = \frac{1}{\bar{X}}[R \cdot Var(\bar{x}_{st}) - Cov(\bar{x}_{st}, \bar{y}_{st})], \tag{5.7.5}$$

where $R = \frac{\bar{Y}}{\bar{X}}$. Now, we know that

$$Var(\bar{x}_{st}) = \sum_{i=1}^{k} w_i^2 \frac{(1-f_i)}{n_i} S_{ix}^2, \tag{5.7.6}$$

where $f_i = \frac{n_i}{N_i}$ and S_{ix}^2 is the mean sum of squares for the character x for the ith stratum, and defined by

$$S_{ix}^2 = \sum_{j=1}^{N_i} \frac{(x_{ij} - \bar{X}_i)^2}{(N_i - 1)}.$$

Next,

$$\begin{aligned} Cov(\bar{x}_{st}, \bar{y}_{st}) &= Cov\left(\sum_{i=1}^{k} w_i \bar{x}_i, \sum_{j=1}^{k} w_i \bar{y}_i\right) \\ &= \sum_{i,j}^{k} w_i w_j Cov(\bar{x}_i, \bar{y}_j) \\ &= \sum_{i=1}^{k} w_i^2 Cov(\bar{x}_i, \bar{y}_i), \end{aligned}$$

since $\bar{x}_i, \bar{y}_j$ are independent for $i \neq j$. Hence, we obtain

$$Cov(\bar{x}_{st}, \bar{y}_{st}) = \sum_{i=1}^{k} w_i^2 \frac{(1 - f_i)}{n_i} S_{ixy}, \tag{5.7.7}$$

where

$$S_{ixy} = \sum_{j=1}^{N_i} \frac{(x_{ij} - \bar{X}_i)(y_{ij} - \bar{Y}_i)}{(N_i - 1)}.$$

Using (5.7.6) and (5.7.7) in (5.7.5), we obtain

$$Bias(\hat{\bar{Y}}_{CR}) = \frac{1}{\bar{X}} \sum_{i=1}^{k} w_i^2 \frac{(1 - f_i)}{n_i} (RS_{ix}^2 - S_{ixy}). \tag{5.7.8}$$

We note that bias of $\hat{\bar{Y}}_{CR}$ decreases as n_i increases.

Theorem 5.7.2. *The approximate variance of the combined ratio estimator $\hat{\bar{Y}}_{CR}$ of the population mean $\bar{Y}$ is given by*

$$Var(\hat{\bar{Y}}_{CR}) = \sum_{i=1}^{k} w_i^2 \frac{(1 - f_i)}{n_i} [S_{iy}^2 + R^2 S_{ix}^2 - 2RS_{ixy}].$$

Proof. We follow the notations of Theorem 5.7.1. The approximate variance of $\hat{\bar{Y}}_{CR}$ is given by

$$\begin{aligned} Var(\hat{\bar{Y}}_{CR}) &= E(\hat{\bar{Y}}_{CR} - \bar{Y})^2 \\ &= \bar{Y}^2 E(-e_1 + e_1^2 + e_2 - e_2 e_1)^2 \\ &= \bar{Y}^2 E(e_1^2 + e_2^2 - 2e_1 e_2), \end{aligned} \tag{5.7.9}$$

retaining terms in (e_1, e_2) of degree two. Using (5.7.3) in (5.7.9), we get

$$Var(\hat{\bar{Y}}_{CR}) = \bar{Y}^2\left[\frac{Var(\bar{x}_{st})}{\bar{X}^2} + \frac{Var(\bar{y}_{st})}{\bar{Y}^2} - \frac{2Cov(\bar{x}_{st}, \bar{y}_{st})}{\bar{X}\bar{Y}}\right]$$
$$= [R^2\, Var(\bar{x}_{st}) + Var(\bar{y}_{st}) - 2R\,Cov(\bar{x}_{st}, \bar{y}_{st})]$$
$$= \sum_{i=1}^{k} w_i^2 \frac{(1-f_i)}{n_i}[S_{iy}^2 + R^2 S_{ix}^2 - 2RS_{ixy}], \quad (5.7.10)$$

which proves the theorem.

We can estimate the variance of $\hat{\bar{Y}}_{CR}$ by

$$Est.Var(\hat{\bar{Y}}_{CR}) = \sum_{i=1}^{k} w_i^2 \frac{(1-f_i)}{n_i}[s_{iy}^2 + \hat{R}^2 s_{ix}^2 - 2\hat{R}s_{ixy}]. \quad (5.7.11)$$

We now derive bias and approximate variance of the separate ratio estimator $\hat{\bar{Y}}_{SR}$ of $\bar{Y}$, the population mean in the following two theorems.

Theorem 5.7.3. *The bias of* $\hat{\bar{Y}}_{SR}$ *is given by*

$$Bias(\hat{\bar{Y}}_{SR}) = \sum_{i=1}^{k} \frac{w_i(1-f_i)}{n_i\bar{X}_i}[R_i S_{ix}^2 - S_{ixy}],$$

where $R_i = \frac{\bar{Y}_i}{\bar{X}_i}$.

Proof. By definition,

$$\hat{\bar{Y}}_{SR} = \sum_{i=1}^{k} w_i \hat{\bar{Y}}_{iR},$$

where $\hat{\bar{Y}}_{iR}$ is the ratio estimator of the ith stratum population mean $\bar{Y}_i$. Using Corollary 5.2.1, the bias of $\hat{\bar{Y}}_{iR}$ is given by

$$Bias(\hat{\bar{Y}}_{iR}) = \frac{1-f_i}{n_i\bar{X}_i}[R_i S_{ix}^2 - S_{ixy}], \quad (5.7.12)$$

where $R_i = \frac{\bar{Y}_i}{\bar{X}_i}$. Hence, bias of $\hat{\bar{Y}}_{SR}$ is obtained as

$$Bias(\hat{\bar{Y}}_{SR}) = \sum_{i=1}^{k} w_i\; Bias(\hat{\bar{Y}}_{iR})$$
$$= \sum_{i=1}^{k} w_i \frac{(1-f_i)}{n_i\bar{X}_i}[R_i S_{ix}^2 - S_{ixy}]. \quad (5.7.13)$$

Thus the theorem is proved.

Remark. If the ratios R_i do not differ very much, that is, if it can be assumed that $R_i = R$, $i = 1, 2, \ldots, k$, then

$$Bias(\hat{\bar{Y}}_{SR}) - Bias(\hat{\bar{Y}}_{CR}) = \sum_{i=1}^{k} \frac{w_i(1-f_i)}{n_i} \left(\frac{1}{\bar{X}_i} - \frac{w_i}{\bar{X}} \right) (RS_{ix}^2 - S_{ixy}).$$

Now $\frac{1}{\bar{X}_i} - \frac{w_i}{\bar{X}} = \frac{\bar{X} - \bar{X}_i w_i}{\bar{X}_i \bar{X}} = \sum_{j \neq i}^{k} \frac{\bar{X}_j w_j}{\bar{X}_i \bar{X}} > 0$, hence assuming $RS_{ix}^2 - S_{ixy} > 0$, bias of separate ratio estimator will be more than that of the combined ratio estimator.

Theorem 5.7.4. *The approximate variance of the separate ratio estimator* $\hat{\bar{Y}}_{SR}$ *is given by*

$$Var(\hat{\bar{Y}}_{SR}) = \sum_{i=1}^{k} \frac{w_i^2(1-f_i)}{n_i} [S_{iy}^2 + R_i^2 S_{ix}^2 - 2R_i S_{ixy}].$$

Proof. We have

$$\hat{\bar{Y}}_{SR} = \sum_{i=1}^{k} w_i \hat{\bar{Y}}_{iR},$$

where $\hat{\bar{Y}}_{iR}$ is the ratio estimator of the ith stratum mean $\bar{Y}_i$. Using Theorem 5.3.1, we obtain the variance of $\hat{\bar{Y}}_{iR}$ as

$$Var(\hat{\bar{Y}}_{iR}) = \frac{1-f_i}{n_i} [S_{iy}^2 + R_i^2 S_{ix}^2 - 2R_i S_{ixy}]. \tag{5.7.14}$$

Hence, we obtain variance of $\hat{\bar{Y}}_{SR}$ as

$$\begin{aligned} Var(\hat{\bar{Y}}_{SR}) &= \sum_{i=1}^{k} w_i^2 Var(\hat{\bar{Y}}_{iR}) \\ &= \sum_{i=1}^{k} \frac{w_i^2(1-f_i)}{n_i} [S_{iy}^2 + R_i^2 S_{ix}^2 - 2R_i S_{ixy}], \end{aligned} \tag{5.7.15}$$

which proves the theorem.

We can estimate $Var(\hat{\bar{Y}}_{SR})$ by

$$Est.Var(\hat{\bar{Y}}_{SR}) \tag{5.7.16}$$

defined by

$$\sum_{i=1}^{k} \frac{w_i^2(1-f_i)}{n_i} [s_{iy}^2 + \hat{R}_i^2 s_{ix}^2 - 2\hat{R}_i s_{ixy}] = \sum_{i=1}^{k} \frac{w_i^2(1-f_i)}{n_i(n_i-1)} \sum_{j}^{n_j} (y_{ij} - \hat{R}_i x_{ij})^2. \tag{5.7.17}$$

Remark. When R_i do not differ from stratum to stratum, that is, if it can be assumed that $R_i = R$, for all i, then $\hat{\bar{Y}}_{CR}$ and $\hat{\bar{Y}}_{SR}$ have the same variance.

5.8 Unbiased Ratio Type Estimator

The unbiased ratio type estimator of the population mean was proposed by Hartley and Ross (1954). Suppose that SRSWOR of n units is selected. Then the unbiased ratio type estimator of $\bar{Y}$ is given by

$$\hat{\bar{Y}}_{HR} = \bar{r}\bar{X} + \frac{n(N-1)}{N(n-1)}(\bar{y} - \bar{r}\bar{x}), \tag{5.8.1}$$

where $\bar{r} = \sum_{i}^{n} \frac{r_i}{n}$, $r_i = \frac{y_i}{x_i}$.

Theorem 5.8.1. *The estimator*

$$\hat{\bar{Y}}_{HR} = \bar{r}\bar{x} + \frac{n(N-1)}{N(n-1)}(\bar{y} - \bar{r}\bar{x})$$

is an unbiased estimator of $\bar{Y}$.

Proof. We have

$$E(\hat{\bar{Y}}_{HR}) = \bar{R}\bar{X} + \frac{n(N-1)}{N(n-1)}[\bar{Y} - E(\bar{r}\bar{x})], \tag{5.8.2}$$

where $\bar{R} = \sum_{i=1}^{N} \frac{r_i}{N}$. Now

$$\begin{aligned} E(\bar{r}\bar{x}) &= Cov(\bar{r}, \bar{x}) + E(\bar{r})E(\bar{x}) \\ &= \frac{N-n}{nN} S_{rx} + \bar{R}\bar{X} \\ &= \frac{N-n}{nN} \cdot \frac{\sum_{i=1}^{N}(r_i - \bar{R})(x_i - \bar{X})}{(N-1)} + \bar{R}\bar{X} \\ &= \frac{(N-n)}{n(N-1)}(\bar{Y} - \bar{R}\bar{X}) + \bar{R}\bar{X} \\ &= \frac{(N-n)\bar{Y}}{n(N-1)} - \frac{N(n-1)}{n(N-1)}\bar{R}\bar{X}. \end{aligned} \tag{5.8.3}$$

Using (5.8.3) in (5.8.2), we get

$$\begin{aligned} E(\hat{\bar{Y}}_{HR}) &= \bar{R}\bar{X} + \frac{n(N-1)}{N(n-1)}\left[\bar{Y} - \frac{(N-n)}{n(N-1)}\bar{Y} - \frac{N(n-1)}{n(N-1)}\bar{R}\bar{X}\right] \\ &= \bar{R}\bar{X} + \bar{Y} - \bar{R}\bar{X} \\ &= \bar{Y}, \end{aligned}$$

which proves the theorem.

A large sample variance of $\hat{\hat{Y}}_{HR}$ is obtained as follows. We assume n and N are large enough so that $\frac{n}{n-1} \cong 1$ and $\frac{N-1}{N} \cong 1$ and take $\bar{r} \cong \bar{R}$. Then

$$\hat{\hat{Y}}_{HR} \cong (\bar{y} - \bar{R}\bar{x}).$$

Hence, large sample variance of $\hat{\hat{Y}}_{HR}$ is given by

$$Var(\hat{\hat{Y}}_{HR}) = \frac{1-f}{n}[S_y^2 + \bar{R}^2 S_x^2 - 2RS_{xy}]. \tag{5.8.4}$$

5.9 Unbiased Ratio Estimator

We consider modification of the sampling design which will make the ratio estimator $\hat{\hat{Y}}_R = \frac{\bar{y}}{\bar{x}}\bar{X}$ unbiased. Consider the Midzuno scheme of sampling discussed in Section 2.5, where the first unit is selected with unequal probabilities of selection and the remaining $(n-1)$ units are drawn with SRSWOR from the $(N-1)$ units. We have seen in Section 2.5, that under this sampling design, the probability of obtaining the sth sample is

$$P(s) = \frac{1}{\binom{N-1}{n-1}} \sum_i^n P_i.$$

If now $P_i \propto x_i$, i.e., $P_i = \frac{x_i}{\sum_1^N x_i} = \frac{x_i}{N\bar{X}}$, then

$$P(s) = \frac{1}{\binom{N-1}{n-1}} \frac{n\bar{x}}{N\bar{X}} = \frac{1}{\binom{N}{n}} \frac{\bar{x}}{\bar{X}}.$$

Hence, under the above design,

$$\begin{aligned} E(\hat{\hat{Y}}_R) &= E\left(\frac{\bar{y}}{\bar{x}}\bar{X}\right) \\ &= \sum_s \frac{\bar{y}}{\bar{x}}\bar{X} \cdot \frac{1}{\binom{N}{n}} \frac{\bar{x}}{\bar{X}} \\ &= \frac{1}{\binom{N}{n}} \sum_s \bar{y} \\ &= \bar{Y}. \end{aligned}$$

5.10 Product Estimator

We have seen in Section 5.4 that ratio estimator is more efficient than the mean of a SRSWOR if $\rho > \frac{1}{2} \cdot \frac{C_x}{C_y}$, if $R > 0$, which is usually the case.

When $\rho < -\frac{1}{2} \cdot \frac{C_x}{C_y}$, there is need of another type of estimator which also makes use of information on auxiliary variable x. Product estimator is an attempt in this direction.

The product estimator of the population mean $\bar{Y}$ is defined as

$$\hat{\bar{Y}}_p = \frac{\bar{y}\bar{x}}{\bar{X}}. \tag{5.10.1}$$

We now obtain bias and variance of $\hat{\bar{Y}}_p$.

(i) Bias of $\hat{\bar{Y}}_p$: We write $\hat{\bar{Y}}_p$ as

$$\begin{aligned}\hat{\bar{Y}}_p = \frac{\bar{y}\bar{x}}{\bar{X}} &= \bar{Y}(1+e_2)(1+e_1)\\ &= \bar{Y}(1+e_1+e_2+e_1e_2),\end{aligned} \tag{5.10.2}$$

where e_1, and e_2 have the same definitions as in (5.2.1). Taking expectation of (5.10.2) and using (5.2.2), we obtain bias of $\hat{\bar{Y}}_p$ as

$$Bias(\hat{\bar{Y}}_p) = \frac{1}{\bar{X}}Cov(\bar{y},\bar{x}) = \frac{1-f}{n\bar{X}}S_{xy}, \tag{5.10.3}$$

which shows that bias of $\hat{\bar{Y}}_p$ decreases as n increases. Bias of $\hat{\bar{Y}}_p$ can be estimated by

$$Est.Bias(\hat{\bar{Y}}_p) = \frac{1-f}{n\bar{X}}s_{xy}. \tag{5.10.4}$$

(ii) Variance of Product Estimator $\hat{\bar{Y}}_p$: Writing $\hat{\bar{Y}}_p$ as in (5.10.2), we find that the variance of the product estimator $\hat{\bar{Y}}_p$ is given by

$$\begin{aligned}Var(\hat{\bar{Y}}_p) &= E(\hat{\bar{Y}}_p - \bar{Y})^2\\ &= \bar{Y}^2E(e_1+e_2+e_1e_2)^2\\ &= \bar{Y}^2E(e_1^2+e_2^2+2e_1e_2),\end{aligned} \tag{5.10.5}$$

Here terms in (e_1, e_2) of degree greater than two are assumed to be negligible. Using (5.2.2), we find that

$$Var(\hat{\bar{Y}}_p) = \frac{1-f}{n}[S_y^2 + R^2S_x^2 + 2RS_{xy}]. \tag{5.10.6}$$

Variance of $\hat{\bar{Y}}_p$ can be estimated by

$$Est.Var(\hat{\bar{Y}}_p) = \frac{1-f}{n}[s_y^2 + \hat{R}^2s_x^2 + 2Rs_{xy}]. \tag{5.10.7}$$

(iii) Comparison with SRSWOR: From the variances of the mean of SRSWOR and the product estimator, we obtain

$$Var(\bar{y})_{SRS} - Var(\hat{\bar{Y}}_p) = -\frac{1-f}{n} RS_x(2\rho S_y + RS_x), \qquad (5.10.8)$$

which shows that $\hat{\bar{Y}}_p$ is more efficient than the simple mean $\bar{y}$ for

$$\rho < -\frac{1}{2} \cdot \frac{C_x}{C_y} \quad \text{if} \quad R > 0$$

and for

$$\rho > \frac{1}{2} \cdot \frac{C_x}{C_y} \quad \text{if} \quad R < 0.$$

5.11 Multivariate Ratio Estimator

Let y be the study variable and $X_1, X_2, \ldots, X_p$ be p auxiliary variates assumed to be correlated with y. Further it is assumed that $X_1, X_2, \ldots, X_p$ are independent. Let $\bar{Y}, \bar{X}_1, \bar{X}_2, \ldots, \bar{X}_p$ be the population means of the variables $y, X_1, X_2, \ldots, X_p$. We assume that a SRSWOR of size n is selected from the population of N units. The following notations will be used.

$S_i^2(s_i^2)$ = the population (sample) mean sum of squares for the variate X_i,
$S_0^2(s_0^2)$ = the population (sample) mean sum of squares for the study variable y.
$C_i = \frac{S_i}{\bar{X}_i}$ = coefficient of variation of the variate X_i,
$C_0 = \frac{S_0}{\bar{Y}}$ = coefficient of variation of the variable y.
$\rho_i = \frac{S_{iy}}{S_i S_0}$ = coefficient of correlation between y and X_i,
$\hat{\bar{Y}}_{Ri} = \frac{\bar{y}}{\bar{x}_i}\bar{X}_i$ = ratio estimator of $\bar{Y}$, based on X_i,

where $i = 1, 2, ..., p$. Then, Olkin (1958) suggested the multivariate ratio estimator of $\bar{Y}$ as follows.

$$\hat{\bar{Y}}_{MR} = \sum_{i=1}^{p} w_i \hat{\bar{Y}}_{Ri}, \quad \sum_{i=1}^{p} w_i = 1$$

$$= \bar{y} \sum_{i=1}^{p} w_i \frac{\bar{X}_i}{\bar{x}_i}. \qquad (5.11.1)$$

(i) Bias of the multivariate rtio estimator: From Corollary 5.2.1, we obtain the bias of $\hat{\bar{Y}}_{Ri}$ as

$$Bias(\hat{\bar{Y}}_{Ri}) = \frac{1-f}{n} \bar{Y}(C_i^2 - \rho_i C_i C_0). \qquad (5.11.2)$$

Using (5.10.2), from (5.10.1), we obtain the bias of $\hat{\hat{Y}}_{MR}$

$$Bias(\hat{\hat{Y}}_{MR}) = \sum_{i=1}^{p} w_i \frac{\bar{Y}(1-f)}{n}(C_i^2 - \rho_i C_i C_0)$$

$$= \frac{\bar{Y}(1-f)}{n} \sum_{i=1}^{p} w_i C_i (C_i - \rho_i C_0). \qquad (5.11.3)$$

(ii) Variance of the multivariate ratio estimator: Using (5.3.4), we obtain the variance of $\hat{\hat{Y}}_{Ri}$ as

$$Var(\hat{\hat{Y}}_{Ri}) = \frac{1-f}{n}\bar{Y}^2(C_0^2 + C_i^2 - 2\rho_i C_0 C_i). \qquad (5.11.4)$$

Using (5.10.4), we obtain from (5.10.1), the variance of $\hat{\hat{Y}}_{MR}$ as

$$Var(\hat{\hat{Y}}_{MR}) = \frac{(1-f)\bar{Y}^2}{n} \cdot \sum_{i=1}^{p} w_i^2 (C_0^2 + C_i^2 - 2\rho_i C_0 C_i). \qquad (5.11.5)$$

EXERCISES

5.1. In a SRSWOR of size n, from a population of N units, the value of y and x are measured, and y and x are their sample means. If X, the population mean of x is known, which of the following estimators will you recommend for estimating $\frac{Y}{X}$? (i) Always use $\frac{y}{X}$, (ii) sometimes $\frac{y}{X}$ and sometimes $\frac{y}{x}$, (iii) always use $\frac{y}{x}$. Give reasons for your answer.

5.2. Table 5.1 gives the data of a hypothetical population of 6 units. Using this data, compare the efficiency of the ratio estimator of the population total Y based on a SRSWOR of two units with that of the usual unbiased estimator by enumerating all possible samples.

5.3. In an experimental study in a field of barley grain, the weight of grain with straw (x) and the grain yield (y) were obtained for a number of sampling units located at random over the field. The total of x is also available. The following data are obtained:

$$C_x^2 = 1.20, \quad C_{xy} = 0.80, \quad C_y^2 = 1.24.$$

Compute the gain in precision obtained by estimating the grain yield of the field from the ratio of grain (y) to total produce (x) instead of from the mean yield of grain per unit.

Table 5.1

x	0	1	3	5	8	10
y	1	3	11	18	29	48

If it requires 20 minutes to cut, thresh and weigh the grain on each unit, 2 minutes to weigh the straw on each unit and 2 hours to collect and weigh the total produce, how many units must be taken per field in order that the ratio estimate may be more economical than the mean per unit?

5.4. If y and x are unbiased estimators of the population totals Y and X, show that the ratio of the exact bias of the ratio estimator $\left(\frac{y}{x}\right)X$ to its standard error is not greater than C_x, the relative standard error of x. Also, show that the bias relative to Y is less than C_x^2 if the relative standard error of $\frac{y}{x}$ is less than C_x.

5.5. If y and x are unbiased estimators of the population totals Y and X, then show that the relative variance of the ratio estimator $\frac{y}{x}$ can be approximated by $C_y^2 - C_x^2$, when the correlation coefficient between $\frac{y}{x}$ and x may be assumed to be negligible.

5.6. Let r_1 and r_2 be estimates of a population ratio at two points of time based on a common probability sample. Show that the bias of the difference $(r_1 - r_2)$ relative to its standard error is less than $2C_{x_2}$ if $\rho(r_1, r_2) \leq 0$, where C_{x_2} is the relative standard error for the estimator of the denominator on the standard error.

5.7. Suppose that a finite population of N units has NP_1 units belonging to a particular category, of which NP_2 units have a special characteristic. It is required to estimate the population ratio $\frac{P_2}{P_1}$ on the basis of a SRSWOR of n units.

(a) If p_1 and p_2 are sample proportions corresponding to p_1 and p_2 respectively, show that $\frac{p_2}{p_1}$ is approximately unbiased for $\frac{P_2}{P_1}$ and derive its approximate variance in case of large samples.

(b) If P_1 is known, prove that the estimator $\frac{P_2}{p_1}$ is more efficient than $\frac{p_2}{p_1}$.

(c) When P_2 is known, derive the condition for $\frac{P_2}{p_1}$ to be more efficient than $\frac{p_2}{p_1}$.

5.8. In small populations the leading term in the bias of $\hat{R}$ in SRSWOR of size n is of the form

$$E(\hat{R} - R) = \frac{(1-f)}{N}b_1 = \frac{b_1}{n} - \frac{b_1}{N},$$

where b_1 does not depend upon n, or N. If $n = mg$ and the sample is divided at random into g groups of size m, let $\hat{R}_i = \frac{\Sigma y}{\Sigma x}$, taken over the remaining $(n-m)$ sample members when the ith group is omitted from the sample. Show that the bias of the estimator

$$w\hat{R} - (w - L)\hat{R}_i$$

both terms in b_1 vanish if $w = g\left[1 - \frac{(n-m)}{N}\right]$.

5.9. Show that for a sample of n units selected with SRSWOR from a population of N units,

$$Bias(\hat{R}_n X) = \frac{n(N-1)}{N(n-1)} E[\bar{x}(\hat{R}_n - \hat{R}_1)],$$

where $\hat{R}_1 = \frac{\bar{y}}{\bar{x}}$, and $\hat{R}_n = \sum_{i=1}^{n} \frac{\left(\frac{y_i}{x_i}\right)}{n}$, $\bar{y}, \bar{x}$ being sample means of y and x. Hence, show that (a) the combined estimator $\theta_1\hat{R}_1\bar{X} + (1-\theta_1)\hat{R}_n\bar{X}$ where θ_1 is a random variable, is unbiased for $\bar{Y}$ if $\theta_1 = \frac{n(N-1)\bar{x}}{N(n-1)\bar{x}}$, and (b) the combined estimator $(1-\theta_2)\hat{R}_1\bar{X} + \theta_2\hat{R}_n\bar{X}$, where θ_2 is a random variable, is almost unbiased if $\theta_2 = -\frac{(N-1)\bar{x}}{N(n-1)\bar{X}}$.

5.10. A sample of size nk is split up at random into k subsamples each of size n. Denote by $Y_R(i)$, $i = 1, 2, \ldots, k$ and $Y_R(nk)$ the usual ratio estimators of the population total based on successive k subsamples of size n and on the sample of size nk respectively. Show that the weighted estimator

$$Y_w = w \sum_{i=1}^{k} Y_R(i) + (1 - kw)Y_R(nk)$$

is unbiased estimator of the population total to the first degree of approximation if

$$w = -\frac{N - nk}{Nk(n-1)}.$$

5.11. Let $\frac{\bar{y}}{\bar{x}}$ be the usual ratio estimator of $R = \frac{\bar{Y}}{\bar{X}}$ in SRSWR. For the exact expression for variance of $\frac{\bar{y}}{\bar{x}}$, show that a sufficient condition for the usual approximation $\frac{Var(\bar{y}-R\bar{x})}{\bar{X}^2}$ to be an understatement for $Var(\frac{\bar{y}}{\bar{x}})$ is that

$$\rho\left[\frac{1}{\bar{x}}, (\bar{y} - R\bar{x})^2\right] \geq 0,$$

where ρ stands for the correlation coefficient.

5.12. Let $y_i,\ x_i\ (i = 1, 2, \ldots, m)$ be unbiased estimators of Y and X respectivelyy, based on m interpreting subsamples of the same size. Prove that $\bar{r}X + \frac{m(\bar{y}-\bar{r}\bar{x})}{(m-1)}$, where $\bar{r} = \sum_{i=1}^{m} \frac{\left(\frac{y_i}{x_i}\right)}{m}$, is an unbiassed estimator of Y.

5.13. The following data are for a small artificial population with $N = 8$ and two strata of equal size.

Stratum 1		Stratum 2	
x_{1i}	y_{1i}	x_{2i}	y_{2i}
2	0	10	7
5	3	18	15
9	7	21	10
15	10	25	16

For a stratified random sample in which $n_1 = n_2 = 2$, compare the MSE's of the separate and the combined ratio estimators by working out the results for all possible samples.

5.14. If the regression of y on x is linear, that is, $E(y|x) = ax+b$, show that in SRSWOR, the estimator $\frac{\bar{y}}{\bar{x}}$ will give a smaller large sample variance than the ratio-type estimator $\bar{r} + \frac{n(N-1)(\bar{y}-\bar{r}\bar{x})}{N(n-1)X}$, where $\bar{r} = \sum_{i=1}^{m} \frac{\left(\frac{y_i}{x_i}\right)}{n}$.

5.15. If the regression between the study variable y and the auxiliary variable x is perfectly linear, that is, $E(y|x_i) = \alpha + \beta x_i,\ i = 1, 2, \ldots, N$, derive the condition for the ratio estimator of the population mean $\bar{Y}$ to be more efficient than the usual unbiased estimator $\bar{y}$ in the case of SRSWRO. What is the effect of the line of regression passing through the origin on the efficiency of the ratio estimator?

5.16. Suppose it is assumed that the population of N units is drawn from the super-population with the following model:

$$E(y|x) = \alpha + \beta x, \qquad Var(y|x) = \sigma^2(1-\rho^2),$$

where $\rho =$ correlation coefficient between y and x. Derive the condition for the ratio estimator to be more efficient than the usual unbiased estimator. From this, derive the condition for the case when the regression line passes through the origin.

5.17. Suppose it is assumed that the population of N units is drawn from the super-population with the following model:

$$\begin{aligned} y_i &= \beta x_i + e_i, \\ E(e_i|x_i) &= 0, \\ E(e_i e_j|x_i, x_j) &= 0 \\ Var(e_i|x_i) &= a x^g, \quad a > 0, \quad g \geq 0. \end{aligned}$$

Show that the ratio estimator of the populatio mean $\bar{Y}$ is more or less efficient than the estimator based on pps and with replacement accordingly as g is less or greater than 1. Show further, that if $g = 1$, the two estimators are equally efficient.

5.18. A SRSWOR $(y'_j, x_{1j}, x_{2j}, \ldots, x_{pj})$, $j = 1, 2, \ldots, n$, is taken from a finite population. The proposed ratio estimator of $\bar{Y}$ is $\bar{y} = \sum_{i=1}^{p} w_i r_i \bar{X}_i$, where $r_i = \frac{\bar{y}'}{\bar{x}_i}$, and $\sum_{i=1}^{p} w_i = 1$. Let $Var(\bar{y}|p)$ and $Var(\bar{y}|p, q)$ denote the variances of $\bar{y}$ based on the auxiliary variables $x_1, x_2, \ldots, x_p$ and $x_1, x_2, \ldots, x_q$, $q > p$. If the allocation of weights w_i is optimum in each case, prove that $Var(\bar{y}|p) \geq Var(\bar{y}|p, q)$.

Chapter 6

REGRESSION METHOD OF ESTIMATION

6.1 Regression Method of Estimation

In Chapter 5, we have seen how to improve upon the conventional estimator $\bar{y}$ by multiplying it with $\frac{\bar{X}}{\bar{x}}$, using the information on auxiliary variate. In this chapter, we shall study another method of improving upon the conventional estimator $\bar{y}$ by utilizing information on the auxiliary variate x. Here, we consider an estimator based on adding to $\bar{y}$, d times the difference $(\bar{x} - \bar{X})$. Thus, we consider a difference estimator

$$\bar{y}_d = \bar{y} + d(\bar{x} - \bar{X}), \tag{6.1.1}$$

where d is some constant. We note that the difference estimator $\bar{y}_d$ is an unbiased estimator of the population mean $\bar{Y}$. Further the variance of $\bar{y}_d$ is given by

$$Var(\bar{y}_d) = Var(\bar{y}) + d^2 Var(\bar{x}) + 2d\, Cov(\bar{y}, \bar{x}). \tag{6.1.2}$$

We select the value of d which minimizes $Var(\bar{y}_d)$. Equating the derivative of (6.1.2) w.r.t. d to zero, we get

$$\begin{aligned} d &= -\frac{Cov(\bar{y}, \bar{x})}{Var(\bar{x})} \\ &= -\frac{\frac{1-f}{n} S_{xy}}{\frac{1-f}{n} S_x^2} = -\beta, \end{aligned} \tag{6.1.3}$$

where β is the regression coefficient of y on x. We assume that the sample is a SRSWOR of size n. Hence, the difference estimator with minimum variance is given by

$$\bar{y}_d = \bar{y} + \beta(\bar{X} - \bar{x}). \tag{6.1.4}$$

The minimum variance of $\bar{y}_d$ is then given by

$$Var(\bar{y}_d)_{min} = Var(\bar{y})(1 - \rho^2), \tag{6.1.5}$$

where ρ is the coefficient of correlation between y and x.

In actual practice β is not known, hence replace β by its estimator, the sample regression coefficient b of y on x. Thus, we get an estimator of $\bar{Y}$ as

$$\bar{y}_r = \bar{y} + b(\bar{X} - \bar{x}), \tag{6.1.6}$$

which we call as the regression estimator of $\bar{Y}$. The estimator b is calculated by

$$b = \frac{\sum_{i=1}^{n}(x_i - \bar{x})(y_i - \bar{y})}{\sum_{i=1}^{n}(x_i - \bar{x})^2} = \frac{s_{xy}}{s_x^2}. \tag{6.1.7}$$

We note that if the regression line of y on x passes through the origin $b = \frac{\bar{y}}{\bar{x}}$, and in this case, the regression estimator becomes the ratio estimator. Thus, if the scatter diagram of the points (y_i, x_i), $i = 1, 2, \ldots, n$ indicates that the line of regression does not pass through the origin, we then use regression estimator.

One common situation in which regression estimator is widely used is that in which one variable is cheaply, quickly and easily measurable while the other one is time-consuming, costly and difficult to measure. The historical example of the use of the regression estimator is in the estimation of the average area of a leaf on a plant by Watson (1937), using weight of a leaf as the auxiliary variate.

6.2 Bias and Variance of the Regression Estimator

The regression estimator of the population mean $\bar{Y}$ is biased. We shall derive its bias and approximate variance in the following two theorems.

Theorem 6.2.1. *The bias of the regression estimator $\bar{y}_r$ is given by*

$$Bias(\bar{y}_r) = -Cov(\bar{x}, b).$$

Proof. We define e, e_1 and e_2 as follows

$$e = \frac{\bar{y} - \bar{Y}}{\bar{Y}}, \quad e_1 = \frac{\bar{x} - \bar{X}}{\bar{X}}, \quad e_2 = \frac{b - \beta}{\beta}. \tag{6.2.1}$$

Then we have

$$\begin{aligned} E(e) &= 0, \quad Var(e) = \tfrac{Var(\bar{y})}{\bar{Y}^2} \\ E(e_1) &= 0, \quad Var(e_1) = \tfrac{Var(\bar{x})}{\bar{X}^2} \\ E(e_2) &= 0, \quad Var(e_2) = \tfrac{Var(b)}{\beta^2} \\ E(e\,e_1) &= \tfrac{Cov(\bar{y},\bar{x})}{\bar{Y}\bar{X}}, \quad E(e_1\,e_2) = \tfrac{Cov(\bar{x},b)}{\bar{X}\beta}. \end{aligned} \tag{6.2.2}$$

Next write $\bar{y}_r$ as

$$\begin{aligned} \bar{y}_r &= \bar{y} + b(\bar{X} - \bar{x}) \\ &= \bar{Y}(1+e) - e_1(1+e_2)\beta\bar{X} \\ &= \bar{Y} + (e\bar{Y} - e_1\beta\bar{X}) - e_1e_2\beta\bar{X}. \end{aligned} \tag{6.2.3}$$

Taking expectation of (6.2.3) and using (6.2.2), we obtain

$$E(\bar{y}_r) = \bar{Y} - Cov(\bar{x}, b), \tag{6.2.4}$$

which proves the theorem.

For large sample, the bias is expected to be small, because usually $Cov(\bar{x}, b)$ will decrease as the sample size increases.

Theorem 6.2.2. *The approximate variance of the regression estimator* $\bar{y}_r$ *is given by*

$$Var(\bar{y}_r) = \frac{1-f}{n} S_y^2(1-\rho^2).$$

Proof. We define e, e_1 and e_2 as in (6.2.1). Then, as in (6.2.3), we write $\bar{y}_r$ as

$$\bar{y}_r = \bar{Y} + (e\bar{Y} - e_1\beta\bar{X}) - e_1e_2\beta\bar{X}. \tag{6.2.5}$$

The approximate variance of $\bar{y}_r$ is obtained as

$$\begin{aligned} Var(\bar{y}_r) &= E(\bar{y}_r - \bar{Y})^2 \\ &= E[(e\bar{Y} - e_1\beta\bar{X}) - e_1e_2\beta\bar{X}]^2 \\ &= E[e^2\bar{Y}^2 - 2e\,e_1\beta\bar{Y}\bar{X} + e_1^2\beta^2\bar{X}^2]. \end{aligned} \tag{6.2.6}$$

Here terms in (e, e_1, e_2) of degree greater than two are assumed to be negligible. Using (6.2.2) in (6.2.6), we obtain

$$\begin{aligned} Var(\bar{y}_r) &= Var(\bar{y}) - 2\beta Cov(\bar{y}, \bar{x}) + \beta^2 Var(\bar{x}) \\ &= Var(\bar{y})(1-\rho^2) \\ &= \frac{1-f}{n} S_y^2(1-\rho^2), \end{aligned} \tag{6.2.7}$$

where ρ is the coefficient of correlation between y and x.

The variance of the regression estimator $\bar{y}_r$ is estimated

$$Est.Var(\bar{y}_r) = \frac{1-f}{n} s_y^2(1-r^2), \tag{6.2.8}$$

where r is the sample coefficient of correlation between y and x.

6.3 Regression Estimator of the Population Total

If it is desired to estimate the population total $Y = N\bar{Y}$, then its regression estimator is obtained as

$$\hat{Y}_r = N\bar{y}_r = N(\bar{y} + b(\bar{X} - \bar{x})).$$

The bias and variance of $\hat{Y}_r$ are obtained by multiplying the bias of $\bar{y}_r$ with N and the variance of $\bar{y}_r$ with N^2 and are given by

$$Bias(\hat{Y}_r) = -N\,Cov(\bar{x}, b) \tag{6.3.1}$$

and

$$Var(\hat{Y}_r) = \frac{N^2(1-f)}{n} S_y^2(1-\rho^2). \tag{6.3.2}$$

Also, the estimator of the variance of $\hat{Y}_R$ is given by

$$Est.Var(\hat{Y}_r) = \frac{N^2(1-f)}{n} s_y^2(1-r^2). \tag{6.3.3}$$

6.4 Comparison with SRSWOR and Ratio Estimator

We compare regression estimator with (i) the sample mean of a SRSWOR, and (ii) regression estimator.

(i) Comparison with SRSWOR: From the variances of $\bar{y}$ and $\bar{y}_r$, we obtain

$$Var(\bar{y}) - Var(\bar{y}_r) = \frac{1-f}{n} S_y^2\rho^2. \tag{6.4.1}$$

Hence we see that when $|\rho| < 1$, regression estimator is more efficient than $\bar{y}$. However, if $\rho = 0$, that is, y and x are uncorrelated, then $\bar{y}_r$ and $\bar{y}$ are equally efficient.

(ii) Comparison with Ratio Estimator: From the variances of $\bar{y}_R$, the ratio estimator and $\bar{y}_r$, we obtain

$$Var(\bar{y}_R) - Var(\bar{y}_r) = \frac{1-f}{n}(RS_x - \rho S_y)^2. \tag{6.4.2}$$

Thus, we see that regression estimator is always more efficient than ratio estimator unless $RS_x - \rho S_y = 0$. If $RS_x - \rho S_y = 0$, then both are equally efficient, which is the situation when the line of regression of y on x is a straight line passing through the origin.

6.5 Regression Estimation in Stratified Sampling

We suppose that the population of N units is divided into k strata and a SRSWOR of size n_i is selected from the ith stratum, $i = 1, 2, \ldots, k$, with $\sum_{i=1}^{k} n_i = n$. As in the case of ratio estimation in stratified sampling, we also have here two regression estimators of the population mean $\bar{Y}$. These are known as (i) a combined regression estimator and (ii) a separate regression estimator.

(i) A Combined Regression Estimator: Here, we consider an estimator of the type

$$\widehat{\bar{Y}}_d = \bar{y}_{st} + d(\bar{x}_{st} - \bar{X}), \tag{6.5.1}$$

where d is a certain constant. We determine d so as to minimize the variance of $\widehat{\bar{Y}}_d$. Thus, taking the variance of $\widehat{\bar{Y}}_d$ and equating its derivative w.r.t. d to zero, we obtain the optimum value of d as

$$d = -\frac{Cov(\bar{y}_{st}, \bar{x}_{st})}{Var(\bar{x}_{st})}. \tag{6.5.2}$$

Now, from (5.7.7), we have

$$Cov(\bar{y}_{st}, \bar{x}_{st}) = \sum_{i=1}^{k} \frac{w_i^2(1-f_i)}{n_i} S_{ixy}, \tag{6.5.3}$$

where $w_i = \frac{N_i}{N}$, $f_i = \frac{n_i}{n}$ and $S_{ixy} = \sum_{j=1}^{N_i} \frac{(y_{ij} - \bar{Y}_i)(x_{ij} - \bar{X}_i)}{(N_i - 1)}$. Also,

$$Var(\bar{x}_{st}) = \sum_{i=1}^{k} \frac{w_i^2(1-f_i)}{n_i} S_{ix}^2, \tag{6.5.4}$$

where w_i and f_i have the same meanings as in (6.5.3) and $S_{ix}^2 = \sum_{j=1}^{N_i} \frac{(x_{ij} - \bar{X}_i)^2}{(N_i - 1)}$. Hence, the optimum value of d is given by

$$d = -\frac{\sum_{i=1}^{k} \frac{w_i^2(1-f_i)}{n_i} S_{ixy}}{\sum_{i=1}^{k} \frac{w_i^2(1-f_i)}{n_i} S_{ix}^2}$$

$$= -\beta, \text{say}$$

where β may be called weighted regression coefficient. Hence, the estimator (6.5.1) becomes

$$\hat{\bar{Y}}_d = \bar{y}_{st} - \beta(\bar{x}_{st} - \bar{X}). \tag{6.5.5}$$

In actual practice, β is not known. Hence β is to be estimated from the sample and estimator of β is provided by

$$b = \frac{\sum_{i=1}^{k} \frac{w_i^2(1-f_i)s_{ixy}}{n_i}}{\sum_{i=1}^{k} \frac{w_i^2(1-f_i)}{n_i} s_{ix}^2}. \tag{6.5.6}$$

Thus, we get a combined regression estimator of $\bar{Y}$ as

$$\hat{\bar{Y}}_{cr} = \bar{y}_{st} + b(\bar{x}_{st} - \bar{X}), \tag{6.5.7}$$

where b is defined by (6.5.6).

For deriving the variance of $\hat{\bar{Y}}_{cr}$, we have the following theorem:

Theorem 6.5.1. *The approximate variance of* $\hat{\bar{Y}}_{cr}$ *is given by*

$$Var(\hat{\bar{Y}}_{cr}) = (1-\rho^2)\sum_{i=1}^{k} \frac{w_i^2(1-f_i)}{n_i} S_{iy}^2$$

where

$$S_{iy}^2 = \sum_{j=1}^{N_i} \frac{(y_{ij} - \bar{Y}_i)^2}{(N_i - 1)}$$

and

$$\rho = \frac{\sum_{i=1}^{k} \frac{w_i^2(1-f_i)}{n_i} S_{ixy}}{\sqrt{\left(\sum_{i=1}^{k} \frac{w_i^2(1-f_i)}{n_i} S_{ix}^2\right)\left(\sum_{i=1}^{k} \frac{w_i^2(1-f_i)}{n_i} S_{iy}^2\right)}}.$$

Proof. We write

$$e = \frac{\bar{y}_{st} - \bar{Y}}{\bar{Y}}, \quad e_1 = \frac{\bar{x}_{st} - \bar{X}}{\bar{X}}, \quad e_2 = \frac{b - \beta}{\beta}. \tag{6.5.8}$$

Then, we have

$$\hat{\bar{Y}}_{cr} = \bar{Y} + (e\bar{Y} - e_1\beta\bar{X}) - \beta\bar{X}e_1e_2$$

and its approximate variance is given by

$$\begin{aligned} Var(\hat{\bar{Y}}_{cr}) &= E(\hat{\bar{Y}}_{cr} - \bar{Y})^2 \\ &= E[(e\bar{Y} - e_1\beta\bar{X}) - \beta\bar{X}e_1e_2]^2 \\ &= E[e^2\bar{Y}^2 - 2e\,e_1\beta\bar{X}\bar{Y} + e_1^2\beta^2\bar{X}^2]. \end{aligned} \tag{6.5.9}$$

Since terms involving (e, e_1, e_2) of degree greater than two are assumed to be negligible. Hence, we obtain

$$Var(\hat{\bar{Y}}_{cr}) = Var(\bar{y}_{st}) - 2\beta Cov(\bar{y}_{st}, \bar{x}_{st}) + \beta^2 Var(\bar{x}_{st}). \tag{6.5.10}$$

We now define

$$\begin{aligned} \rho &= \frac{Cov(\bar{y}_{st}, \bar{x}_{st})}{\sqrt{Var(\bar{y}_{st}) \cdot Var(\bar{x}_{st})}} \\ &= \frac{\sum_{i=1}^{k} \frac{w_i^2(1-f_i)}{n_i} S_{ixy}}{\sqrt{\left(\sum_{i=1}^{k} \frac{w_i^2(1-f_i)}{n_i} S_{iy}^2\right)\left(\sum_{i=1}^{k} \frac{w_i^2(1-f_i)}{n_i} S_{ix}^2\right)}}. \end{aligned}$$

Note that $\beta = \frac{Cov(\bar{y}_{st}, \bar{x}_{st})}{Var(\bar{x}_{st})} = \rho\sqrt{\frac{Var(\bar{y}_{st})}{Var(\bar{x}_{st})}}$. Using this, we can write (6.5.10) as

$$\begin{aligned} Var(\hat{\bar{Y}}_{cr}) &= (1-\rho^2) Var(\bar{y}_{st}) \\ &= (1-\rho^2) \sum_{i=1}^{k} \frac{w_i^2(1-f_i)}{n_i} S_{iy}^2. \end{aligned} \tag{6.5.11}$$

We can estimate the variance of $\hat{\bar{Y}}_{cr}$ by

$$Est.Var(\hat{\bar{Y}}_{cr}) = (1-r^2) \sum_{i=1}^{k} \frac{w_i^2(1-f_i)}{n_i} s_{iy}^2, \tag{6.5.12}$$

where

$$r = \frac{\sum_{i=1}^{k} \frac{w_i^2(1-f_i)}{n_i} s_{ixy}}{\sqrt{\left(\sum_{i=1}^{k} \frac{w_i^2(1-f_i)}{n_i} s_{iy}^2\right)\left(\sum_{i=1}^{k} \frac{w_i^2(1-f_i)}{n_i} s_{ix}^2\right)}}.$$

(ii) Separate Regression Estimator: For the ith stratum, the regression estimator is

$$\bar{y}_{ir} = \bar{y}_i + b_i(\bar{x}_i - \bar{X}_i), \quad i = 1, 2, \ldots, k \tag{6.5.13}$$

where

$$b_i = \frac{\sum_{j=1}^{n_i}(y_{ij} - \bar{y}_i)(x_{ij} - \bar{x}_i)}{\sum_{j=1}^{n_i}(x_{ij} - \bar{x}_i)^2}.$$

Then, the separate regression estimator is defined as

$$\begin{aligned}\widehat{\bar{Y}}_{sr} &= \sum_{i=1}^{k} w_i \bar{y}_{ir} \\ &= \sum_{i=1}^{k} w_i\{\bar{y}_i + b_i(\bar{x}_i - \bar{X}_i)\}.\end{aligned} \tag{6.5.14}$$

Now, from (6.3.2), we have

$$Var(\bar{y}_{ir}) = \frac{1 - f_i}{n_i} S_{iy}^2(1 - \rho_i^2),$$

where

$$\rho_i = \frac{\sum_{j=1}^{N_i}(y_{ij} - \bar{Y}_i)(x_{ij} - \bar{X}_i)}{\sqrt{\sum_{j}^{N_i}(y_{ij} - \bar{Y}_i)^2 \cdot \sum_{j=1}^{N_i}(x_{ij} - \bar{X}_i)}}. \tag{6.5.15}$$

Hence, variance of $\widehat{\bar{Y}}_{sr}$ is obtained as

$$\begin{aligned}Var(\widehat{\bar{Y}}_{sr}) &= \sum_{i=1}^{k} w_i^2 \cdot Var(\bar{y}_{ir}) \\ &= \sum_{i=1}^{k} \frac{w_i^2(1 - f_i)(1 - \rho_i^2)}{n_i} S_{iy}^2.\end{aligned} \tag{6.5.16}$$

Thus, we have proved the following theorem.

Theorem 6.5.2. *The approximate variance of $\widehat{\bar{Y}}_{sr}$ is given by*

$$Var(\widehat{\bar{Y}}_{sr}) = \sum_{i=1}^{k} \frac{w_i^2(1 - f_i)(1 - \rho_i^2)}{n_i} S_{iy}^2.$$

We can estimate the variance of $\widehat{\bar{y}}_{sr}$ by

$$Est.Var(\widehat{\bar{Y}}_{sr}) = \sum_{i=1}^{k} \frac{w_i^2(1 - f_i)(1 - r_i^2)s_{iy}^2}{n_i}, \tag{6.5.17}$$

where

$$r_i = \frac{\sum_{j=1}^{n_i}(y_{ij} - \bar{y}_i)(x_{ij} - \bar{x}_i)}{\sqrt{\sum_{j=1}^{n_i}(y_{ij} - \bar{y}_i)^2 \cdot \sum_{j=1}^{n_i}(x_{ij} - \bar{x}_i)^2}}.$$

Remark. If ρ_i do not differ from stratum to stratum, that is, if $\rho_i = \rho$, $i = 1, 2, \ldots, k$, then the variances of the combined regression estimator and the separate regression estimator are same.

6.6 Multivariate Regression Estimator

Let y be the study variable and $X_1, X_2, \ldots, X_p$ be p auxiliary variates. A SRSWOR of size n is selected and thus n observations on each of the variables $y, X_1, X_2, \ldots, X_p$ are obtained. Let $\bar{y}, \bar{x}_1, \bar{x}_2, \ldots, \bar{x}_p$ be the sample means, and $\bar{Y}, \bar{X}_1, \ldots, \bar{X}_p$ be the population means. Let b_i be the regression coefficient of y on the variable X_i, defined by

$$b_i = \frac{\sum_{j=1}^{n}(y_j - \bar{y})(x_{ij} - \bar{x}_i)}{\sum_{j=1}^{n}(x_{ij} - \bar{x}_i)^2}, \quad i = 1, 2, \ldots, p.$$

Then, Ghosh (1947) suggested a multivariate estimator of the form

$$\hat{\bar{Y}}_{Mr} = \bar{y} + \sum_{i=1}^{p} b_i(\bar{x}_i - \bar{X}_i). \tag{6.6.1}$$

Des Raj (1965) suggested a weighted difference estimator of the form

$$\hat{\bar{Y}}_{Md} = \sum_{i=1}^{p} w_i\{\bar{y}_i + d_i(\bar{x}_i - \bar{X}_i)\}, \tag{6.6.2}$$

where the weights w_i add up to 1 and d_i are known constants. Usually d_i are taken as $\frac{\bar{Y}}{\bar{X}_i}$, the values of which are obtained from past surveys.

EXERCISES

6.1. An experienced farmer makes an eye estimate of the weight of apples (x), on each tree in an apple field of $N = 100$ trees and finds a total weight $X = 550$ kg. A SRSWOR of $n = 10$ trees is selected and appless are picked and weighed with the following data.

Tree No.	1	2	3	4	5	6	7	8	9	10
Actual Wt (y)	30	26	21	35	25	28	29	20	33	22
Ast. Wt. (x)	29	30	24	38	26	29	30	22	34	24

Compute the regression estimator of the total weight of apples and show that the difference estimator is more efficient?

6.2. A rough measurement x, made on each unit, is related to the true measurement y on the unit by the equation $x = y + e + d$, where d is a constant bias and e is an error of measurement which is uncorrelated with y, and has mean o and variance S_e^2 in the population, assumed infinite. In SRSWOR of size n, compare the variance of (i) the difference estimator $[\bar{y} + (\bar{X} - \bar{x})]$ of the mean $\bar{Y}$ and (ii) the linear regression estimator.

6.3. For the hypothetical population of 6 units given in Table 5.1, obtain the relative standard error of the regression estimator of the population total Y based on a SRSWOR of 2 units by enumerating all possible samples and compare its efficiency with those of the ratio estimator and the usual unbiased estimator.

6.4. Suppose it is assumed that the population of N units is drawn from a super-population with the following model:

$$E(y_i|x_i) = \alpha + \beta x_i,$$
$$Var(y_i|x_i) = a x_i^g, \quad a > 0, \quad g \geq 0,$$
$$Cov(y_i, y_j|x_i, x_j) = 0, \quad i \neq j.$$

A SRSWOR of size n is selected from this popualtion. Show that the regression estimator of the population mean is always more efficient than the ratio estimator when x is used as the auxiliary variable.

6.5. Show that if the proportional increase in the variance of the regression estimator arising due to the use of a constant λ instead of β is less than α, then the relative deviation of λ from β, that is $\frac{(\lambda-\beta)}{\beta}$ must be less than $\sqrt{\frac{\alpha(1-\rho^2)}{\rho^2}}$, where β is the regression coefficient of y on x and ρ is the correlation coefficient between y and x.

6.6. By working out all possible samples of size 2 selected from each stratum with SRSWOR, compare the MSE's of the separate and combined regression estimators of the population total Y of the population of 8 units divided into 2 strata.

Stratum 1		Stratum 2	
y	x	y	x
0	4	7	5
3	6	12	6
5	7	13	8

Chapter 7

CLUSTER SAMPLING

7.1 Cluster Sampling

When the list of elements of the population is not available, it is not possible to use element as a sampling unit. In this situation, if the list of groups of elements is available, then the group of elements can be used as a sampling unit. The groups of elements in which the population is divided are called clusters. Cluster sampling is a sampling in which a cluster is taken as a sampling unit. In cluster sampling a sample of clusters is selected and then information on all the elements of the selected clusters is obtained. For instance, suppose we wish to select a sample of persons residing in a city. Now, the list of persons residing in the city is not readily available, so we cannot use person as a sampling unit. Here, it may be possible to have a list of houses. So, house is used as a sampling unit. We select a sample of houses and then enumerate all the persons of the selected houses. Here, house is a cluster of persons.

It is necessary that no element of the population should belong to more than one cluster. Each element of the population must belong to one and only one cluster. Cluster sampling is operationally easier and also cheaper.

7.2 Estimation in Case of Equal Clusters

We shall consider the case when clusters are of equal size. Suppose that the population is divided into N clusters, each cluster having M elements. We use the following notations:

y_{ij} = value of the jth unit in the ith cluster, $j = 1, 2, \ldots, M$;
$i = 1, 2, \ldots, N$.

$$\bar{y}_{i\cdot} = \frac{1}{M}\sum_{j=1}^{M} y_{ij} = \text{mean of the } i^{th} \text{ cluster, } i = 1, 2, \ldots, N.$$

$$\bar{y}_{\cdot\cdot} = \frac{1}{NM}\sum_{i=1}^{N}\sum_{j=1}^{M} y_{ij} = \text{mean per element of the population}$$

$$= \frac{1}{N}\sum_{i=1}^{N} \bar{y}_{i\cdot} = \text{mean of cluster means.}$$

$$S_i^2 = \frac{1}{M-1}\sum_{j=1}^{M}(y_{ij} - \bar{y}_{i\cdot})^2$$

$$= \text{mean sum of squares for the } i^{th} \text{ cluster, } i = 1, 2, \ldots, N.$$

$$\bar{S}_w^2 = \frac{1}{N}\sum_{i=1}^{N} S_i^2$$

$$= \text{mean sum of squares within clusters in the population.}$$

$$S_b^2 = \frac{1}{N-1}\sum_{i=1}^{N}(\bar{y}_i - \bar{y}_{\cdot\cdot})^2$$

$$= \text{mean sum of squares between cluster means in the population.}$$

$$S^2 = \frac{1}{NM-1}\sum_{i=1}^{N}\sum_{j=1}^{M}(\bar{y}_{ij} - \bar{y}_{\cdot\cdot})^2$$

$$= \text{mean sum of squares for the whole population.}$$

We assume that a simple random sample of n clusters is selected without replacement and all the nM elements of these n clusters are enumerated. Let

$$\bar{y}_c = \frac{1}{n}\sum_{i=1}^{n} \bar{y}_{i.} = \frac{1}{nM}\sum_{i=1}^{n}\sum_{j=1}^{M} y_{ij}$$

$$= \text{sample mean of cluster means}$$

$$= \text{mean per element in the sample}$$

$$s_b^2 = \frac{1}{(n-1)}\sum_{i=1}^{n}(\bar{y}_{i.} - \bar{y}_c)^2$$

$$= \text{mean sum of squares between cluster means in the sample}$$

$$\bar{s}_w^2 = \frac{1}{n}\sum_{i=1}^{n} S_i^2$$

= mean sum of squares within clusters in the sample

Theorem 7.2.1.

(i) $\bar{y}_c$ *is an unbiased estimator of the population mean* $\bar{y}_{..}$.
(ii) $Var(\bar{y}_c) = \left(\frac{1}{n} - \frac{1}{N}\right) S_b^2$.

Proof. $\bar{y}_c$ is the mean of n means $\bar{y}_{i\cdot}$ drawn from a population of N means $\bar{y}_{i\cdot}$, $i = 1, 2, \ldots, N$ with SRSWOR. Hence, from the theory of SRSWOR, we get

$$E(\bar{y}_c) = \frac{1}{N}\sum_{i=1}^{N} \bar{y}_{i\cdot} = \bar{y}_{..}, \tag{7.2.1}$$

and

$$\begin{aligned} Var(\bar{y}_c) &= \left(\frac{1}{n} - \frac{1}{N}\right) \frac{\sum_{i=1}^{N}(\bar{y}_{i\cdot} - \bar{y}_{..})^2}{N-1} \\ &= \left(\frac{1}{N} - \frac{1}{N}\right) S_b^2. \end{aligned} \tag{7.2.2}$$

In the following theorem, we shall derive the variance of $\bar{y}_c$ in terms of ρ, the intraclass correlation between pairs of elements of clusters.

Theorem 7.2.2. *The variance of* $\bar{y}_c$ *is given by*

$$Var(\bar{y}_c) = \frac{(N-n)}{N} \cdot \frac{(NM-1)}{M(N-1)} \cdot \frac{S^2}{nM}\{1 + (M-1)\rho\},$$

where

$$\begin{aligned} \rho &= \frac{\dfrac{\sum_{i=1}^{N}\sum_{j \neq k}^{M}(y_{ij} - \bar{y}_{..})(y_{ik} - \bar{y}_{..})}{NM(M-1)}}{\dfrac{\sum_{i=1}^{N}\sum_{j=1}^{M}(y_{ij} - \bar{y}_{..})^2}{NM}} \\ &= \frac{\sum_{i=1}^{N}\sum_{j \neq k}^{M}(y_{ij} - \bar{y}_{..})(y_{ik} - \bar{y}_{..})}{(NM-1)(M-1)S^2}. \end{aligned}$$

Proof. We have

$$\begin{aligned}
&\sum_{i=1}^{N}\sum_{j\neq k}^{M}(y_{ij}-\bar{y}_{..})(y_{ik}-\bar{y}_{..}) \\
&= \sum_{i=1}^{N}\left[\sum_{j=1}^{M}\sum_{k=1}^{M}(y_{ij}-\bar{y}_{..})(y_{ik}-\bar{y}_{..}) - \sum_{j=1}^{M}(y_{ij}-\bar{y}_{..})^2\right] \\
&= \sum_{i=1}^{N}\left[M^2(\bar{y}_{i\cdot}-\bar{y}_{..})^2 - \sum_{j=1}^{M}(y_{ij}-\bar{y}_{i\cdot}+\bar{y}_{i\cdot}-\bar{y}_{..})^2\right] \\
&= \sum_{i=1}^{N}\left[M^2(\bar{y}_{i\cdot}-\bar{y}_{..})^2 - \left(\sum_{j=1}^{M}(y_{ij}-\bar{y}_{i\cdot})^2 + M(\bar{y}_{i\cdot}-\bar{y}_{..})^2\right)\right] \\
&= \sum_{i=1}^{N} M(M-1)(\bar{y}_{i\cdot}-\bar{y}_{..})^2 - (M-1)S_i^2 \\
&= M(M-1)(N-1)S_b^2 - N(M-1)\bar{S}_w^2. \qquad (7.2.3)
\end{aligned}$$

Hence, the value of ρ becomes

$$\rho = \frac{M(N-1)S_b^2 - N\bar{S}_w^2}{(NM-1)S^2}. \qquad (7.2.4)$$

Now, we have the identity

$$\sum_{i=1}^{N}\sum_{j=1}^{M}(y_{ij}-\bar{y}_{..})^2 = \sum_{i=1}^{N}\sum_{j=1}^{M}(y_{ij}-\bar{y}_{i\cdot})^2 + \sum_{i=1}^{N}\sum_{j=1}^{M}(\bar{y}_{i\cdot}-\bar{y}_{..})^2,$$

that is,

$$(NM-1)S^2\rho = N(M-1)\bar{S}_w^2 + M(N-1)S_b^2. \qquad (7.2.5)$$

From (7.2.4), we have

$$(NM-1)S^2 = -N\bar{S}_w^2 + M(N-1)S_b^2. \qquad (7.2.6)$$

Eliminating $\bar{S}_w^2$ between (7.2.5) and (7.2.6), we obtain S_b^2 as

$$S_b^2 = \frac{(NM-1)S^2}{M^2(N-1)}[1+(M-1)\rho]. \qquad (7.2.7)$$

Now, from (7.2.2) and (7.2.7), we have

$$Var(\bar{y}_c) = \frac{N-n}{Nn}S_b^2 = \frac{N-n}{N}\cdot\frac{(NM-1)}{M(N-1)}\cdot\frac{S^2}{nM}[1+(M-1)\rho]. \qquad (7.2.8)$$

If N is very large, we obtain

$$Var(\bar{y}_c) \cong \frac{S^2}{nM}[1 + (M-1)\rho]. \quad (7.2.9)$$

Remark. If $M = 1$, then,

$$Var(\bar{y}_c) = \frac{N-n}{nN}S^2,$$

which is the variance of the mean of a SRSWOR of size n, and in this case the cluster sample is the SRSWOR. When $M > 1$, usually $\rho > 0$, and hence the sampling variance increases with the size of the cluster.

7.3 Efficiency of Cluster Sampling

The variance of the sample mean of cluster means is given by

$$Var(\bar{y}_c) = \frac{N-n}{Nn}S_b^2. \quad (7.3.1)$$

Now, if an equivalent simple random sample of nM elements is drawn without replacement, then

$$\begin{aligned} Var(\bar{y})_{SRS} &= \frac{NM - nM}{NM \cdot nM}S^2 \\ &= \frac{N-n}{NnM}S^2. \end{aligned} \quad (7.3.2)$$

Hence, efficiency of cluster sampling relative to an equivalent SRSWOR is obtained as

$$E = \frac{Var(\bar{y})_{SRS}}{Var(\bar{y}_c)} = \frac{S^2}{MS_b^2}. \quad (7.3.3)$$

From the identity

$$\sum_{i=1}^{N}\sum_{j=1}^{M}(y_{ij} - \bar{y}_{..})^2 = \sum_{i=1}^{N}\sum_{j=1}^{M}(y_{ij} - \bar{y}_{i.})^2 + \sum_{i=1}^{N}\sum_{j=1}^{M}(\bar{y}_{i.} - \bar{y}_{..})^2, \quad (7.3.4)$$

we can set up the analysis of variance as given in Table 7.1.

From the analysis of variance Table 7.1, we see that the efficiency of cluster sampling relative to an equivalent SRSWOR is given by the ratio of total mean sum of squares to the mean sum of squares between the clusters.

We now derive the efficiency of the cluster sampling relative to an equivalent SRSWOR in terms ρ, the intraclass correlation coefficient.

Table 7.1 ANOVA table for the population.

Source	SS	d.f.	MSS
Between Clusters	$M\sum_{i=1}^{N}(\bar{y}_{i\cdot}-\bar{y}_{\cdot\cdot})^2$	$N-1$	MS_b^2
Within Clusters	$\sum_{i=1}^{N}\sum_{j=1}^{M}(y_{ij}-\bar{y}_{i\cdot})^2$	$N(M-1)$	$\bar{S}_w^2$
Total	$\sum_{i=1}^{N}\sum_{j=1}^{M}(y_{ij}-\bar{y}_{\cdot\cdot})^2$	$NM-1$	S^2

Theorem 7.3.1. *The efficiency of cluster sampling relative to an equivalent SRSWOR is given by*

$$E = \frac{M(N-1)}{(NM-1)}[1+(M-1)\rho]^{-1}.$$

Proof. The efficiency of cluster sampling relative to an equivalent SRSWOR is obtained by dividing (7.3.2) by (7.2.8) and is given by

$$E = \frac{M(N-1)}{(NM-1)} \cdot \frac{1}{[1+(M-1)\rho]},$$

which proves the result.

If N is very large then the efficiency E becomes

$$E \cong \frac{1}{[1+(M-1)\rho]}. \tag{7.3.6}$$

7.4 Estimation of Efficiency

Since $\bar{y} = \frac{1}{n}\sum_{i=1}^{n}\bar{y}_{i\cdot}$ is the mean of n means $\bar{y}_{i\cdot}$ from a population of N means $\bar{y}_{i\cdot}$, $i = 1, 2, \ldots, N$ drawn with SRSWOR, it follows from theory of SRSWOR that

$$E\left[\frac{1}{n-1}\sum_{i=1}^{n}(\bar{y}_{i\cdot}-\bar{y}_c)^2\right] = \frac{1}{N-1}\sum_{i=1}^{N}(\bar{y}_{i\cdot}-\bar{y}_{\cdot\cdot})^2$$
$$= S_b^2. \tag{7.4.1}$$

Hence, s_b^2 is an unbiased estimator of S_b^2. Further, $\bar{s}_w^2 = \frac{1}{n}\sum_{i=1}^{n} S_i^2$ is the mean n mean sum of squares S_i^2 drawn from the population of N mean

Table 7.2 ANOVA table for the sample.

Source	SS	d.f.	MSS
Between Clusters	$M\sum_{i=1}^{n}(\bar{y}_{i\cdot}-\bar{y}_c)^2$	$n-1$	Ms_b^2
Within Clusters	$\sum_{i=1}^{n}\sum_{j=1}^{M}(y_{ij}-\bar{y}_{i\cdot})^2$	$n(M-1)$	$\bar{s}_w^2$
Total	$\sum_{i=1}^{n}\sum_{j=1}^{M}(y_{ij}-\bar{y}_c)^2$	$nM-1$	

sums of squares S_i^2, $i=1,2,\ldots,N$, hence it follows from the theory of SRSWOR that

$$E(\bar{s}_w^2) = E\left(\frac{1}{n}\sum_{i=1}^{n} S_i^2\right) = \frac{1}{N}\sum_{i=1}^{N} S_i^2 = \bar{S}_w^2. \tag{7.4.2}$$

Thus, $\bar{s}_w^2$ is an unbiased estimator of $\bar{S}_w^2$. We now show how to estimate S^2 unbiasedly. From (7.2.5), we have

$$(NM-1)S^2 = N(M-1)\bar{S}_w^2 + M(N-1)S_b^2,$$

from which we obtain an unbiased estimator of S^2 as

$$\hat{S}^2 = \frac{1}{(NM-1)}[N(M-1)\bar{s}_w^2 + M(N-1)\bar{s}_b^2]. \tag{7.4.3}$$

Hence, an estimate of efficiency $E = \frac{S^2}{MS_b^2}$ is obtained as

$$\hat{E} = \frac{[N(M-1)\bar{s}_w^2 + M(N-1)s_b^2]}{M(NM-1)s_b^2}. \tag{7.4.4}$$

From the following identity for the sample,

$$\sum_{i=1}^{n}\sum_{j=1}^{M}(y_{ij}-\bar{y}_c)^2 = \sum_{i=1}^{n}\sum_{j=1}^{M}(y_{ij}-\bar{y}_{i\cdot})^2 + \sum_{i=1}^{n}\sum_{j=1}^{M}(\bar{y}_{i\cdot}-\bar{y}_c)^2$$

we have

$$n(M-1)s^2 = n(M-1)\bar{s}_w^2 + M(n-1)s_b^2$$

and we can set up the analysis of variance as in Table 7.2.

When N is very large, $\hat{E}$ is given by

$$\begin{aligned}\hat{E} &\cong \frac{(M-1)\bar{s}_w^2 + Ms_b^2}{M^2 s_b^2} \\ &= \frac{1}{M} + \frac{(M-1)}{M}\cdot\frac{\bar{s}_w^2}{Ms_b^2}.\end{aligned} \tag{7.4.5}$$

7.5 Optimum Size of Cluster Sampling

From the formula of the sampling variance in cluster sampling, we see that for a given total sample size in terms of units, the sampling variance increases with cluster size M and decreases with increasing number n of clusters. We must remember that the cost of the survey decreases with increasing cluster size and increases with increasing number of clusters. Hence, it is necessary to compromise between these two opposing points of view by determining optimum values of the cluster size M and the number n of the clusters, which would minimize the sampling variance for a fixed cost or minimize the cost for a specified sampling variance.

We shall assume the following cost function.

$$C = c_1 nM + c_2\sqrt{n}, \tag{7.5.1}$$

where c_1 is the cost of enumeration and processing the information per element and c_2 is the cost proportional to distance between the clusters.

Mahalanobis (1946) and Jessen (1942), on the basis of extensive empirical studies, have shown the following relationship between $\bar{S}_w^2$ and M:

$$\bar{S}_w^2 = aM^b, \tag{7.5.2}$$

where a and b are positive constants. Then, from (7.2.5), one obtains

$$S_b^2 = \frac{(NM-1)S^2 - N(M-1)aM^b}{M(N-1)}. \tag{7.5.3}$$

If N is large, then, from (7.5.3), we obtain

$$S_b^2 = S^2 - a(M-1)M^{b-1}, \tag{7.5.4}$$

and the variance of cluster sampling is given by

$$\begin{aligned} V = Var(\bar{y}_c) &= \frac{1}{n}S_b^2 \\ &= \frac{1}{n}[S^2 - a(M-1)M^{b-1}]. \end{aligned}$$

We shall find the optimum values of n and M which would minimize the sampling variance for fixed cost C_0, say. Thus, we minimize $V = Var(\bar{y}_c)$ subject to $C_0 = c_1 nM + c_2\sqrt{n}$, that is, we minimize

$$\phi = V + \lambda[c_1 nM + c_2\sqrt{n} - C_0], \tag{7.5.6}$$

with respect to n and M, where λ is a Lagrangian multiplier. Equating the partial derivatives of ϕ w.r.t. n and M to zero we obtain

$$\frac{\partial\phi}{\partial n} = \frac{\partial V}{\partial n} + \lambda\left(c_1 M + \frac{c_2}{2\sqrt{n}}\right) = 0 \tag{7.5.7}$$

$$\frac{\partial \phi}{\partial M} = \frac{\partial V}{\partial M} + \lambda c_1 n = 0. \tag{7.5.8}$$

Now from (7.5.5), we obtain

$$\frac{\partial V}{\partial n} = -\frac{1}{n} V, \tag{7.5.9}$$

and from (7.5.8), we obtain

$$\lambda = -\frac{V}{M} \cdot \frac{1}{nc_1}. \tag{7.5.10}$$

Substituting from (7.5.9) and (7.5.10) in (7.5.7), we obtain

$$\frac{1}{V} \cdot \frac{\partial V}{\partial M} = -\frac{2c_1\sqrt{n}}{2\sqrt{n}c_1 M + c_2}. \tag{7.5.11}$$

Treating $C_0 = c_1 n M + c_2 \sqrt{n}$ as a quadratic equation in $\sqrt{n}$, one obtains

$$\sqrt{n} = \frac{-c_2 + \sqrt{c_2^2 + 4c_1 M C_0}}{2c_1 M}.$$

Substituting this value of $\sqrt{n}$ in (7.5.11), we obtain

$$\frac{M}{V} \frac{\partial V}{\partial M} = -1 + \left(1 + \frac{4c_1 M C_0}{c_2^2}\right)^{-\frac{1}{2}}. \tag{7.5.12}$$

Now, from (7.5.5), we find that

$$\frac{\partial V}{\partial M} = \frac{-a[M^{b-1} + (M-1)(b-1)M^{b-2}]}{n},$$

from which on multiplication by M and dividing by n, one obtains

$$\frac{M}{V} \frac{\partial V}{\partial M} = -\frac{a[M^b + (M-1)(b-1)M^{b-1}]}{S^2 - a(M-1)M^{b-1}}. \tag{7.5.13}$$

Equating (7.5.12) and (7.5.13), we have an equation for determinating the optimum value of m as

$$\frac{a[M^b + (M-1)(b-1)M^{b-1}]}{S^2 - a(M-1)M^{b-1}} = 1 - \left(1 + \frac{4c_1 M C_0}{c_2^2}\right)^{-\frac{1}{2}} \tag{7.5.14}$$

which is to be solved by trial and error method. Denoting the optimum value of M by $\hat{M}$, the optimum value of n is obtained from (7.5.12) as

$$\sqrt{\hat{n}} = \frac{-c_2 + \sqrt{c_2^2 + 4c_1 \hat{M} C_0}}{2c_1 \hat{M}}. \tag{7.5.15}$$

We note that from (7.5.5) that as M increases V decreases, hence $\frac{M}{V}\frac{\partial V}{\partial M}$ remains nearly constant. Hence from (7.5.12), we see that $\frac{c_1 M C_0}{c_2^2}$ remains constant.

7.6 Estimation of a Proportion

In this section, we consider the problem of estimation of P, the proportion of units in the population having a specified attribute on the basis of a sample of clusters.

Suppose that a sample of n clusters is drawn with SRSWOR. Defining $y_{ij} = 1$ if the jth unit in the ith cluster belongs to the specified category (i.e. possessing the given attribute), we find that

$$\bar{y}_{i\cdot} = P_i, \qquad \bar{y}_{\cdot\cdot} = \frac{1}{N}\sum_{i=1}^{N} P_i = P,$$

$$S_i^2 = \frac{MP_iQ_i}{(M-1)}, \qquad \bar{S}_w^2 = \frac{M\sum_{i=1}^{N} P_iQ_i}{N(M-1)}, \qquad S^2 = \frac{NMPQ}{NM-1},$$

$$\begin{aligned} S_b^2 &= \frac{1}{N-1}\sum_{i=1}^{N}(P_i - P)^2 \\ &= \frac{1}{N-1}\left[\sum_{i=1}^{N} P_i^2 - NP^2\right] \\ &= \frac{1}{(N-1)}\left[-\sum_{i=1}^{N} P_i(1-P_i) + \sum_{i=1}^{N} P_i - NP^2\right] \\ &= \frac{1}{(N-1)}\left[NPQ - \sum_{i=1}^{N} P_iQ_i\right], \end{aligned}$$

where P_i is the proportion of elements in the ith cluster, belonging to the specified category and $Q_i = 1 - P_i$, $i = 1, 2, \ldots, N$ and $Q = 1 - P$. Then, applying Theorem 7.2.1, we find that

$$\hat{P}_c = \frac{1}{n}\sum_{i=1}^{n} P_i \tag{7.6.1}$$

is an unbiased estimator of P and

$$Var(\hat{P}_c) = \frac{N-n}{nN} \cdot \frac{\left[NPQ - \sum_{i=1}^{N} P_iQ_i\right]}{(N-1)}. \tag{7.6.2}$$

This variance of $\hat{P}_c$ can b expressed as

$$Var(\hat{P}_c) = \frac{N-n}{N-1}\frac{PQ}{nM}[1 + (M-1)\rho], \tag{7.6.3}$$

where the value of ρ can be obtained from (7.2.4) on substitution for S_b^2, $\bar{S}_w^2$ and S^2 as

$$\rho = 1 - \frac{M}{(M-1)} \cdot \frac{1}{N} \frac{\sum_{i=1}^{N} P_i Q_i}{PQ}. \tag{7.6.4}$$

The variance of $\hat{P}_c$ can be estimated unbiasedly by

$$\begin{aligned} Est.Var(\hat{P}_c) &= \frac{N-n}{nN} s_b^2 \\ &= \frac{N-n}{nN} \cdot \frac{1}{(n-1)} \sum_{i=1}^{n} (P_i - \hat{P}_c)^2 \\ &= \frac{N-n}{Nn(n-1)} \left[n\hat{P}_c\hat{Q}_c - \sum_{i}^{n} P_i Q_i \right] \end{aligned} \tag{7.6.5}$$

where $\hat{Q}_c = I - \hat{P}_c$. Using Theorem 7.3.1, the efficiency of cluster sampling relative to SRSWOR is given by

$$E = \frac{(N-1)}{NM-1} \cdot \frac{NPQ}{\left(NPQ - \sum_{i=1}^{N} P_i Q_i \right)}. \tag{7.6.6}$$

If N is large, then $E \cong \frac{1}{M}$.

An estimator of the total number of elements belonging to a specified category is obtained by multiplying $\hat{P}_c$ by NM, i.e. by $NM\hat{P}_c$. The expressions of variance and variance estimator are obtained by multiplying the corresponding expressions for $\hat{P}_c$ by N^2M^2.

7.7 Estimation in Case of Unequal Clusters

Let there be N clusters and let M_i be the number of elements in the ith cluster, $i = 1, 2, \ldots, N$. Let $M_0 = \sum_{i=1}^{N} M_i$, and $\bar{M} = \frac{1}{N} \sum_{i=1}^{N} M_i = \frac{M_0}{N}$. Then, the population mean is given by

$$\bar{y}_{..} = \frac{\sum_{i=1}^{N} \sum_{j=1}^{M} y_{ij}}{N\bar{M}} = \frac{\sum_{i=1}^{N} M_i \bar{y}_{i.}}{N\bar{M}}.$$

Suppose that n clusters are selected with SRSWOR and all elements in these selected clusters are surveyed. We assume here that M_i, $i = 1, 2, \ldots, N$, are known.

Theorem 7.7.1. *An unbiased estimator of the population mean is given by*

$$\hat{\bar{Y}}_c = \frac{1}{n}\sum_{i=1}^{n}\left(\frac{M_i}{\bar{M}}\right)\bar{y}_{i\cdot}$$

with its variance given by

$$Var(\hat{\bar{Y}}_c) = \frac{N-n}{Nn}S_b'^2,$$

where

$$S_b'^2 = \frac{1}{N-1}\sum_{i=1}^{N}\left(\frac{M_i\bar{y}_{i\cdot}}{\bar{M}} - y_{\cdot\cdot}\right)^2.$$

Proof. $\hat{Y}_c$ is the arithmetic mean of n values $\frac{M_i\bar{y}_i}{M}$ drawn with SRSWOR from a population of N values $\frac{M_i\bar{y}_i}{M}$. Hence, using the theory SRSWOR, we obtain

$$\begin{aligned}
E(\hat{\bar{Y}}_c) &= E\left(\frac{1}{n}\sum_{i=1}^{n}\frac{M_i}{\bar{M}}\bar{y}_{i\cdot}\right)\\
&= \frac{1}{N}\sum_{i=1}^{N}\frac{M_i}{\bar{M}}\bar{y}_{i\cdot}\\
&= \bar{y}_{\cdot\cdot}, \qquad (7.7.1)
\end{aligned}$$

and

$$\begin{aligned}
Var(\hat{\bar{Y}}_c) &= \frac{N-n}{nN}\frac{\sum_{i=1}^{N}\left(\frac{M_i}{\bar{M}}\bar{y}_{i\cdot} - \bar{y}_{\cdot\cdot}\right)^2}{N-1}\\
&= \frac{N-n}{nN}S_b'^2, \qquad (7.7.2)
\end{aligned}$$

which proves the theorem.

Using the theory of SRSWOR, we see that an unbiased estimator of $Var(\hat{\bar{Y}}_c)$ is given by

$$Est.Var(\hat{\bar{Y}}_c) = \frac{N-n}{Nn}\cdot s_b'^2, \qquad (7.7.3)$$

where

$$s_b'^2 = \frac{1}{(n-1)} \sum_{i=1}^{n} \left(\frac{M_i \bar{y}_{i\cdot}}{\bar{M}} - \hat{\bar{Y}}_c \right)^2 .$$

We now consider estimators when the values of M_i are known only for the sample clusters and not for all the clusters.

Theorem 7.7.2. *An unbiased estimator of the population total is given by*

$$\hat{Y} = \frac{N}{n} \sum_{i=1}^{n} M_i \bar{y}_{i\cdot}$$

with variance given by

$$Var(\hat{Y}) = \frac{N-n}{Nn} \frac{\sum_{i=1}^{N} (NM_i \bar{y}_{i\cdot} - Y)^2}{N-1}.$$

Proof. $\hat{Y}$ is the arithmetic mean of a sample of n values $NM_i\bar{y}_{i\cdot}$ drawn with SRSWOR from a population of N values $NM_i\bar{y}_{i\cdot}$. Hence, using the theory of SRSWOR, we get

$$\begin{aligned} E(\hat{Y}) &= E\left[\frac{1}{n} \sum_{i=1}^{n} NM_i \bar{y}_{i\cdot}\right] \\ &= \frac{1}{N} \sum_{i=1}^{N} NM_i \bar{y}_{i\cdot} \\ &= \sum_{i=1}^{N} M_i \bar{y}_{i\cdot} = Y \end{aligned} \tag{7.7.4}$$

and

$$Var(\hat{Y}) = \frac{N-n}{Nn} \cdot \frac{\sum_{i=1}^{N} (NM_i \bar{y}_{i\cdot} - Y)^2}{N-1}. \tag{7.7.5}$$

An unbiased estimator of the variance of $\hat{Y}$ is obtained as

$$Est.Var(\hat{Y}) = \frac{N-n}{Nn} \cdot \frac{\sum_{i=1}^{n} (NM_i \bar{y}_{i\cdot} - \hat{Y})^2}{(n-1)}. \tag{7.7.6}$$

In the above, note that although we can estimate the total unbiasedly, we cannot estimate the population mean since $\sum_{i=1}^{N} M_i$ is not known.

We now give below estimators of the population mean when M_i's are known only for the sample clusters.

(i) $\bar{Y}'_c = \frac{1}{n}\sum_{i=1}^{n} \bar{y}_{i\cdot}$.

(ii) $\bar{Y}'' = \dfrac{\sum_{i=1}^{n} M_i \bar{y}_{i\cdot}}{\sum_{i=1}^{n} M_{i\cdot}}$.

We study their biases and mean square errors.

(i) $\bar{Y}'_c$: We have

$$E(\bar{Y}'_c) = E\left[\frac{1}{n}\sum_{i=1}^{n} \bar{y}_{i\cdot}\right]$$
$$= \frac{1}{N}\sum_{i=1}^{N} \bar{y}_{i\cdot} \neq \bar{y}_{\cdot\cdot}\,.$$

Hence, $\bar{Y}'_c$ is a biassed estimator of the $\bar{y}_{\cdot\cdot}$, the population mean. The bias of $\bar{Y}'_c$ is obtained as

$$Bias(\bar{Y}'_c) = E(\bar{Y}'_c) - \bar{y}_{\cdot\cdot}$$
$$= \frac{1}{N}\sum_{i=1}^{N} \bar{y}_{\cdot\cdot} - \bar{y}_{\cdot\cdot}$$
$$= \frac{1}{N}\sum_{i=1}^{N}\sum_{j=1}^{M_i} \frac{y_{ij}}{M_i} - \frac{\sum_{i=1}^{N}\sum_{j=1}^{M_j} y_{ij}}{N\bar{M}}$$
$$= \frac{1}{N}\sum_{i=1}^{N}\sum_{j=1}^{M_i} \left(\frac{1}{M_i} - \frac{1}{\bar{M}}\right) y_{ij}$$
$$= -\frac{1}{N\bar{M}}\sum_{i=1}^{N}\sum_{j=1}^{M_i} (M_i - \bar{M})\frac{y_{ij}}{M_i}$$
$$= -\frac{1}{N\bar{M}}\sum_{i=1}^{N} (M_i - \bar{M})\bar{y}_{i\cdot}$$
$$= -\frac{1}{N\bar{M}}\sum_{i=1}^{N} (M_i - \bar{M})(\bar{y}_{i\cdot} - \bar{\bar{y}}_N)$$
$$= -\frac{N-1}{M_0} S_{w\bar{y}}$$

where $\bar{\bar{y}}_N = \frac{1}{N}\sum_{i=1}^{N} \bar{y}_{i\cdot}$ and

$$S_{w\bar{y}} = \frac{\sum_{i=1}^{N}(M_i - \bar{M})(\bar{y}_{i\cdot} - \bar{\bar{y}}_N)}{(N-1)} = Cov(M_i, \bar{y}_{i\cdot}).$$

Next we obtain the mean square error of $\bar{Y}_c'$.

$$MSE(\bar{Y}_c') = Var\left(\frac{1}{n}\sum^{n} \bar{y}_{i\cdot}\right) + [Bias]^2$$

$$= \left(\frac{1}{n} - \frac{1}{N}\right) S_b^2 + \left(\frac{N-1}{M_0}\right)^2 S_{w\bar{y}}^2,$$

where $S_b^2 = \frac{\sum_{i=1}^{N}(\bar{y}_{i\cdot} - \bar{\bar{y}}_N)^2}{N-1}$.

(ii) $\bar{Y}_c''$: We note that $\bar{Y}_c'' = \frac{\sum_{i=1}^{n} M_i\bar{y}_{i\cdot}}{\sum_{i=1}^{n} M_i}$ can be considered as a ratio estimator $\frac{\bar{u}}{\bar{x}}$ by considering $u_i = M_i\bar{y}_{i\cdot}$ and $x_i = M_i$. Hence, the results (5.5.1) and (5.5.3) of Chapter 5 can be applied to obtain an approximate variance of $\bar{Y}_c''$. We leave the derivation as an exercise to the reader.

7.8 pps with Replacement

In many practical situations, the cluster total for the study variable is likely to be positively correlated with the number of units in the cluster. In this situation, it is advantageous to select clusters with probability proportional to the number of units in the cluster instead of with equal probability, or to stratify the clusters according to their sizes and then to draw a SRSWOR of clusters from each stratum. We consider here the case where clusters are selected with probability proportional to the number of units in the cluster and with replacement.

Suppose that n clusters are selected with ppswr, the size being the number of units in the cluster. Here, P_i, the probability of selection assigned to the ith cluster is given by

$$P_i = \frac{M_i}{M_0} = \frac{M_i}{N\bar{M}}, \quad i = 1, 2, \ldots, N. \tag{7.8.1}$$

Theorem 7.8.1. *An unbiased estimator of the population mean is given by*

$$\hat{\bar{Y}}_c = \frac{1}{n}\sum_{i=1}^{n} \bar{y}_{i\cdot}\ .$$

Proof. We have

$$\begin{aligned}\hat{\bar{Y}}_c &= \frac{1}{n}\sum_{i=1}^{n} \bar{y}_{i\cdot} \\ &= \frac{1}{n}\sum_{i=1}^{N} a_i \bar{y}_{i\cdot},\end{aligned}$$

where a_i denotes the number of times the ith cluster occurs in the sample. The random variables $a_1, a_2, \ldots, a_N$ follow a multinomial probability distribution with

$$\begin{aligned} E(a_i) = nP_i, \quad Var(a_i) = nP_i(1 - P_i) \\ Cov(a_i, a_j) = -nP_iP_j, \quad i \neq j. \end{aligned} \tag{7.8.2}$$

Hence,

$$\begin{aligned} E(\hat{\bar{Y}}_c) &= \frac{1}{n}\sum_{i=1}^{N} E(a_i)\bar{y}_{i\cdot} \\ &= \frac{1}{n}\sum_{i=1}^{N} nP_i\bar{y}_{i\cdot} \\ &= \sum_{i=1}^{N} \frac{M_i}{N\bar{M}}\bar{y}_{i\cdot} \\ &= \frac{\sum_{i=1}^{N}\sum_{j=1}^{M_i} y_{ij}}{N\bar{M}} = \bar{y}_{\cdot\cdot}, \end{aligned}$$

which proves the result.

We now derive the variance of $\hat{\bar{Y}}_c$.

Theorem 7.8.2. *The variance of* $\hat{\bar{Y}}_c$ *is given by*

$$Var(\hat{\bar{Y}}_c) = \frac{1}{nN\bar{M}}\sum_{i=1}^{N} M(\bar{y}_{i\cdot} - \bar{y}_{\cdot\cdot})^2.$$

Proof. From $\hat{\bar{Y}}_c = \frac{1}{n}\sum_{i=1}^{N} a_i \bar{y}_{i\cdot}$, where a_i's are as defined in the proof of Theorem 7.8.1, we obtain

$$\begin{aligned}
Var(\hat{\bar{Y}}_c) &= \frac{1}{n^2}\left[\sum_{i=1}^{N} Var(a_i)\bar{y}_{i\cdot}^2 + \sum_{i\neq j}^{N} Cov(a_i, a_j)\bar{y}_{i\cdot}\bar{y}_{j\cdot}\right] \\
&= \frac{1}{n}\left[\sum_{i=1}^{N} P_i(1-P_i)\bar{y}_{i\cdot}^2 - \sum_{i\neq j}^{N} P_i P_j \bar{y}_{i\cdot}\bar{y}_{j\cdot}\right] \\
&= \frac{1}{n}\left[\sum_{i=1}^{N} P_i \bar{y}_{i\cdot}^2 - \left(\sum_{i=1}^{N} P_i \bar{y}_{i\cdot}\right)^2\right] \\
&= \frac{1}{n}\sum_{i=1}^{N} P_i(\bar{y}_{i\cdot} - \bar{y}_{\cdot\cdot})^2 \\
&= \frac{1}{nN\bar{M}}\sum_{i=1}^{N} M_i(\bar{y}_{i\cdot} - \bar{y}_{\cdot\cdot})^2, \qquad (7.8.3)
\end{aligned}$$

which proves the theorem.

Theorem 7.8.3. *An unbiased estimator of the variance of* $\hat{\bar{Y}}_c$ *is given by*

$$Est.Var(\hat{\bar{Y}}_c) = \frac{1}{n(n-1)}\sum_{i=1}^{n}(\bar{y}_{i\cdot} - \hat{\bar{Y}}_c)^2.$$

Proof. We have

$$\begin{aligned}
E&\left[\frac{1}{n(n-1)}\sum_{i=1}^{n}(\bar{y}_{i\cdot} - \hat{\bar{Y}}_c)^2\right] \\
&= E\left[\frac{1}{n(n-1)}\left(\sum_{i=1}^{n}\bar{y}_{i\cdot}^2 - n\hat{\bar{Y}}_c^2\right)\right] \\
&= \frac{1}{n(n-1)}\left[E\left(\sum_{1}^{N} a_i \bar{y}_{i\cdot}^2\right) - n\,Var(\hat{\bar{Y}}_c) - n\bar{y}_{\cdot\cdot}^2\right],
\end{aligned}$$

where a_i's are as defined in (7.8.2). Hence

$$
\begin{aligned}
E&\left[\frac{1}{n(n-1)}\sum_{i=1}^{n}(\bar{y}_{i\cdot}-\hat{\bar{Y}}_c)^2\right]\\
&=\frac{1}{n(n-1)}\left[\sum_{i=1}^{N}nP_iy_{i\cdot}^2-n\cdot\frac{1}{n}\sum_{i=1}^{N}P_i(\bar{y}_{i\cdot}-\bar{y}_{\cdot\cdot})^2-n\bar{y}_{\cdot\cdot}^2\right]\\
&=\frac{1}{(n-1)}\left[\sum_{i=1}^{N}P_i(\bar{y}_{i\cdot}^2-\bar{y}_{\cdot\cdot}^2)-\frac{1}{n}\sum_{i=1}^{N}P_i(\bar{y}_{i\cdot}-\bar{y}_{\cdot\cdot})^2\right]\\
&=\frac{1}{(n-1)}\left[\sum_{i=1}^{N}P_i(\bar{y}_{i\cdot}-\bar{y}_{\cdot\cdot})^2-\frac{1}{n}P_i(\bar{y}_{i\cdot}-\bar{y}_{\cdot\cdot})^2\right]\\
&=\frac{1}{n}\sum_{i=1}^{N}P_i(\bar{y}_{i\cdot}-\bar{y}_{\cdot\cdot})^2\\
&=Var(\hat{\bar{Y}}_c),
\end{aligned}
$$

which proves the theorem.

EXERCISES

7.1. A population consisting of 2500 elements is divided into 10 strata, each containing 50 large units composed of 5 elements. The analysis of variance for a study variable y for the population is given in Table 7.3. Is the relative precision of the large to the small unit greater with SRSWOR than with stratified simple random sampling with proportional allocation? (Ignore fpc.)

7.2. In cluster sampling, each cluster contains M units. The cost of the survey is given by

$$C = 4tMn + 60\sqrt{n},$$

where t is the time in hours spent getting answers from a single element. If $2000 is spent on the survey, determine the values of n for

Table 7.3 Analysis of variance table.

Source	d.f.	Mean SS
Between strata	9	30.6
Between large units within strata	490	3.0
Between elements within large units	2000	1.6

$M = 1, 5, 10$, and $t = \frac{1}{2}$ hr, 2 hrs., if the variance of the sample mean (ignoring fpc) is

$$\frac{S^2}{Mn}[1 + (M-1)\rho].$$

7.3. In Exercise 7.2, determine the optimum size of the cluster which would minimize the variance for fixed cost of \$5000 in each case (a) $t = \frac{1}{2}$ hr, and (b) $t = 2$ hrs.

7.4. A population consists of N clusters, each consisting of M units. Let P_i be the proportion of units in the ith cluster possessing a certain characteristic and P the corresponding proportion for the whole population. If a SRSWOR of n clusters is selected, show that $\hat{P} = \frac{1}{n}\sum_{i=1}^{n} P$ is an unbiased estimator of P and that its variance is given by

$$Var(\hat{P}) = \left(\frac{1}{n} - \frac{1}{N}\right)\frac{1}{(N-1)}\sum_{i=1}^{N}(P_i - P)^2.$$

If an equivalent sample of nM units is selected with SRSWOR from NM units, give the corresponding estimator anad its variance, and discuss the efficiency of cluster sampling relative to this SRSWOR of nM units.

7.5. In cluster sampling, a sample of n clusters of M units each was selected with SRSWR. Let b and w be unbiasesd estimates of between-cluster and within-cluster variances. Assuming the sample size in terms of the number of units to be fixed, obtain an estimate of the relative efficiency of cluster sampling as compared to that of direct sampling of units by estimating sampling variances in the two cases unbiasedly.

7.6. For examining the efficiency of sampling households instead of persons for estimating the proportion of females in a given region, the following assumptions are made (i) each household consists of 4 persons (husband, wife, and 2 children) and (ii) sex of a child is binomially distributed. Show that the intraclass correlation coefficient in this case is $-\frac{1}{6}$ and that the efficiency of sampling households compared to that of sampling persons is 200%.

7.7. Suppose a population consists of N clusters of M units each. When n clusters are selected systematically for estimating the population mean per unit, derive the sampling variance of the estimator in terms of the intraclass correlation coefficient ρ between pairs of units in the clusters and the ρ' the intraclass correlation coefficient between pairs of clusters in the samples, assuming N to be a multiple of n.

7.8. If NM units of the population are grouped at random to form N clusters of M units each, show that sampling n clusters with SRSWOR would have the same efficiency as sampling nM units with SRSWOR.

7.9. A population consists of N clusters, M_i being the size of the ith cluster, $i = 1, 2, \ldots, N$. Clusters are selected one by one with replacement and with *pp* to the size of the cluster. The cluster selected at the $(r+1)^{th}$ draw is rejected if the number of distinct clusters selected in the first r draws is a pre-assigned number n. Let the ith cluster occur r_i times in the sample of r_i draws, $r = 0, 1, 2, \ldots,\ i = 1, 2, \ldots, N$. If $\bar{y}_{i\cdot}$ is the mean of the ith cluster, show that $y = \sum_{i=1}^{N} \frac{r_i}{r} \bar{y}_{i\cdot}$ is an unbiased estimator of the population mean. Further, show that an unbiased estimator of the variance of this estimator is given by

$$Est.Var(y) = \frac{\sum_{i=1}^{N} r_i(\bar{y}_{i\cdot} - \bar{Y})^2}{r(r-1)}.$$

7.10. A population consists of N clusters, the ith cluster consisting of M_i units. Let ρ' denote the intraclass correlation coefficient defined by

$$\rho' = \frac{\sum_{i=1}^{N} \sum_{j \neq k}^{M_i} (y_{ij} - \bar{y}_{\cdot\cdot})(y_{ik} - \bar{y}_{\cdot\cdot})}{M_0(\bar{M} - 1)S^2},$$

where $M_0 = \sum_{i=1}^{N} M_i$, $\bar{M} = \frac{M_0}{N}$, and S^2 is the mean sum of squares of the population.

(i) Show that

$$\frac{1}{M-1} < -\frac{M_0 - 1}{M(M-1)} \leq \rho' \leq \frac{N-1}{N} \frac{S_b^{-2}}{S^2},$$

where

$$S_b^{-2} = \frac{\sum_{i=1}^{N} M_i(M_i - 1)(\bar{y}_{i\cdot} - \bar{y}_{\cdot\cdot})^2}{(N-1)M(M-1)}.$$

(ii) If clusters are of equal size, then show that

$$S_{\bar{b}}^{-2} = S_b^2 = \text{the mean sum of squares between cluster means in the population,}$$

$$\rho' = \frac{NM}{NM-1}\rho \cong \rho, \text{ where}$$

$M_i = M$, $i = 1, 2, \ldots, N$, and ρ is defined by

$$\rho = \frac{\sum_{i=1}^{N} \sum_{j \neq k}^{M} (y_{ij} - \bar{y}_{..})(y_{ik} - \bar{y}_{..})}{(NM-1)(M-1)S^2}.$$

7.11. The population is divided into N clusters, the ith cluster being of size M_i. Consider the following estimator of the population mean $\bar{y}_{..}$.

$$\bar{y}^* = \frac{\sum_{i=1}^{n} M_i \bar{y}_{i.}}{\sum_{i=1}^{n} M_i}.$$

Show that the variance of $\bar{y}^*$ is given approximately by

$$Var(\bar{y}^*) \cong \frac{N-n}{N} \cdot \frac{S^2}{n\bar{M}}[1 + (\bar{M}-1)\rho'],$$

where ρ', $\bar{M}$ and S are as defined in Exercise 7.10.

Chapter 8

SUB-SAMPLING TWO-STAGE AND THREE-STAGE SAMPLING

8.1 Sub-sampling

In cluster sampling discussed in Chapter 7, we have seen that all the elements in the selected clusters are surveyed. Also, we observed that in cluster sampling, the sampling variance increased with larger clusters and thus it became less efficient than that when an element is used as a sampling unit. Thus, we see that for given number of elements, greater precision wil be obtained by distributing them over a large number of clusters than by taking a small number of clusters and enumerating all elements within them. Sub-sampling is an attempt in this direction. In sub-sampling, a sample of clusters is first selected and then from each selected cluster, another sample of specified number of elements is selected. The clusters which form units of sampling at the first stage are called first-stage units, while the elements within the clusters which form sampling units at the second stage are called second-stage units. This sampling is also known as two-stage sampling. This sampling procedure can easily be extended to three-stage sampling or multi-stage sampling. Consider crop surveys in which villages are first-stage units, fields within villages are second-stage units and plots within fields are the third-stage units. This is an example of three-stage sampling.

8.2 Two-stage Sampling — Equal First-stage Units

We shall assume that the population consists of NM elements grouped into N first-stage units of M second-stage units each. Suppose a sample of n first-stage units is selected and a sample of m second-stage units is selected from each selected first-stage unit. Further, we assume that units at each stage are selected with SRSWOR. We use the following notations:

y_{ij} = the value of the jth second-stage unit in the ith first-stage unit, $j = 1, 2, \ldots, M;\ i = 1, 2, \ldots, N.$

$$\bar{y}_{i\cdot} = \frac{1}{M}\sum_{j=1}^{M} y_{ij}$$

= mean per second-stage unit in the ith first-stage unit, $i = 1, 2, \ldots, N$.

$$\bar{y}_{\cdot\cdot} = \frac{\sum_{i=1}^{N}\sum_{j=1}^{M} y_{ij}}{NM}$$

= mean per second-stage unit in the population

$$\bar{y}_{im} = \frac{1}{m}\sum_{j=1}^{m} y_{ij}$$

= mean per second-stage unit in the ith first-stage unit in the sample.

$$\bar{y}_{nm} = \frac{\sum_{i=1}^{n}\sum_{j=1}^{m} y_{ij}}{nm} = \frac{1}{n}\sum_{i=1}^{n} \bar{y}_{im}$$

= mean per second-stage unit in the sample.

Theorem 8.2.1.

(i) $\bar{y}_{nm}$ *is an unbiased estimator of the population mean.*

(ii) $Var(\bar{y}_{nm}) = \left(\frac{1}{n} - \frac{1}{N}\right) S_b^2 + \frac{1}{n}\left(\frac{1}{m} - \frac{1}{M}\right) \bar{S}_w^2$, *where*

$$S_b^2 = \frac{1}{(N-1)}\sum_{i=1}^{N}(\bar{y}_{i\cdot} - \bar{y}_{\cdot\cdot})^2$$

$$\bar{S}_w^2 = \frac{1}{N}\sum_{i=1}^{N} S_i^2$$

$$= \frac{1}{N}\sum_{i=1}^{N}\frac{1}{(M-1)}\sum_{j=1}^{M}(y_{ij} - \bar{y}_{i\cdot})^2$$

$$= \frac{1}{N(M-1)}\sum_{i=1}^{N}\sum_{j=1}^{M}(y_{ij} - \bar{y}_{i\cdot})^2$$

Proof. Since the sample is selected in two stages, the expected value is worked out in two stages. Let E_2 denote the expected value over all possible samples of size m from each of n fixed first-stage units and E_1 denote the expected value over all possible samples of size n and similarly define the variances V_1 and V_2. Thus,

$$\begin{aligned}
E[\bar{y}_{nm}] &= E\left[\frac{1}{n}\sum_{i=1}^{N}\bar{y}_{im}\right] \\
&= E_1\left[\frac{1}{n}\sum_{i=1}^{n}E_2(\bar{y}_{im}|i)\right] \\
&= E_1\left[\frac{1}{n}\sum_{i=1}^{n}\bar{y}_{i\cdot}\right] \\
&= \frac{1}{N}\sum_{i=1}^{N}\bar{y}_{i\cdot} = \bar{y}_{\cdot\cdot}, \qquad (8.2.1)
\end{aligned}$$

which proves the first part of the theorem.

Next,

$$\begin{aligned}
Var(\bar{y}_{nm}) &= Var\left(\frac{1}{n}\sum_{i=1}^{n}\bar{y}_{im}\right) \\
&= V_1E_2\left(\frac{1}{n}\sum_{i=1}^{n}\bar{y}_{im}|i\right) + E_1V_2\left(\frac{1}{n}\sum_{i=1}^{n}\bar{y}_{im}|i\right) \\
&= V_1\left(\frac{1}{n}\sum_{i=1}^{n}\bar{y}_{i\cdot}\right) + E_2\left[\frac{1}{n^2}\sum_{i=1}^{n}\left(\frac{1}{m}-\frac{1}{M}\right)S_i^2\right] \\
&= \left(\frac{1}{n}-\frac{1}{N}\right)S_b^2 + \frac{1}{n}\left(\frac{1}{m}-\frac{1}{M}\right)\cdot\frac{1}{N}\sum_{i=1}^{N}S_i^2 \\
&= \left(\frac{1}{n}-\frac{1}{N}\right)S_b^2 + \frac{1}{n}\left(\frac{1}{m}-\frac{1}{M}\right)\bar{S}_w^2. \qquad (8.2.2)
\end{aligned}$$

Remark 1. The variance of $\bar{y}_{nm}$ consists of two components. If the selected first-stage unit had been completely surveyed, that is, $m = M$, then the variance of the sample mean would be given by the first component only. But, actually, each selected first-stage unit is partially surveyed by means of a sample of m second-stage units drawn from it. Hence, the second term of (8.2.2) represents the contribution to the sampling variance arising from sub-sampling the selected first-stage units.

Let us consider the case of $n = N$. This corresponds to the case when every first-stage unit is sampled. The sampling variance is then given by the second term of (8.2.2). But, actually n first-stage units are sampled, hence the first term in (8.2.2) represents the contribution arising from sampling the first-stage units. The case $n = N$ can be looked upon as stratified sampling with first-stage units as strata. Thus, two-stage sampling can be considered as incomplete stratified sampling.

Remark 2. Limiting Cases:

(i) When N is large relative to n, then

$$Var(\bar{y}_{nm}) = \frac{S_b^2}{n} + \frac{1}{n}\left(\frac{1}{m} - \frac{1}{M}\right)\bar{S}^2. \tag{8.2.3}$$

(ii) When M is large relative to m, then

$$Var(\bar{y}_{nm}) = \left(\frac{1}{n} - \frac{1}{N}\right) S_b^2 + \frac{\bar{S}_w^2}{nm}. \tag{8.2.4}$$

(iii) When N and M are both large relative to n and m respectively, then

$$Var(\bar{y}_{nm}) = \frac{S_b^2}{n} + \frac{\bar{S}_w^2}{nm}. \tag{8.2.5}$$

We shall now show how to estimate the variance of $\bar{y}_{nm}$. This is done in Theorem 8.2.2.

Theorem 8.2.2. *An unbiased estimator of the variance of $\bar{y}_{nm}$ is given by*

$$Est.Var(\bar{y}_{nm}) = \left(\frac{1}{n} - \frac{1}{N}\right) s_b^2 + \frac{1}{N}\left(\frac{1}{m} - \frac{1}{M}\right)\bar{s}_w^2,$$

where

$$s_b^2 = \frac{1}{(n-1)}\sum_{i=1}^{n}(\bar{y}_{im} - \bar{y}_{nm})^2$$

$$s_i^2 = \frac{1}{(m-1)}\sum_{j=1}^{m}(y_{ij} - \bar{y}_{im})^2$$

$$\bar{s}_w^2 = \frac{1}{n}\sum_{i=1}^{n} s_i^2.$$

Proof. We have from the definition of s_b^2,

$$(n-1)s_b^2 = \sum_{i=1}^{n} \bar{y}_{im}^2 - n\bar{y}_{nm}^2. \tag{8.2.6}$$

Hence,

$$\begin{aligned}
(n-1)E(s_b^2) &= \sum_{i=1}^{n} E(\bar{y}_{im}^2) - nE(\bar{y}_{nm}^2) \\
&= \sum_{i=1}^{n} E_1\left(E_2(\bar{y}_{im}^2|i)\right) - n(Var(\bar{y}_{nm}) + \bar{y}_{..}^2) \\
&= \sum_{i=1}^{n} E_1[Var(\bar{y}_{im}) + \bar{y}_{i.}^2] \\
&\quad - n\left[\left(\frac{1}{n} - \frac{1}{N}\right) S_b^2 + \frac{1}{n}\left(\frac{1}{m} - \frac{1}{M}\right)\bar{S}_w^2\right] - n\bar{y}_{..}^2 \\
&= \sum_{i=1}^{n} E_1\left[\left(\frac{1}{m} - \frac{1}{M}\right) S_i^2 + \bar{y}_{i.}^2\right] \\
&\quad - n\left[\left(\frac{1}{n} - \frac{1}{N}\right) S_b^2 + \frac{1}{n}\left(\frac{1}{m} - \frac{1}{M}\right)\bar{S}_w^2\right] - n\bar{y}_{..}^2 \\
&= n \cdot \left(\frac{1}{m} - \frac{1}{M}\right)\frac{1}{N}\sum_{i=1}^{N} S_i^2 + n \cdot \frac{1}{N}\sum_{i=1}^{N}\bar{y}_{i.}^2 \\
&\quad - n\left[\left(\frac{1}{n} - \frac{1}{N}\right) S_b^2 + \frac{1}{n}\left(\frac{1}{m} - \frac{1}{M}\right)\bar{S}_w^2\right] - n\bar{y}_{..}^2 \\
&= n\left(\frac{1}{m} - \frac{1}{M}\right)\bar{S}_w^2 + \frac{n}{N}\sum_{i=1}^{N}(\bar{y}_{i.} - \bar{y}_{..})^2 \\
&\quad - n\left[\left(\frac{1}{n} - \frac{1}{N}\right) S_b^2 + \frac{1}{n}\left(\frac{1}{m} - \frac{1}{M}\right)\bar{S}_w^2\right] \\
&= (n-1)\left(\frac{1}{m} - \frac{1}{M}\right)\bar{S}_w^2 + \frac{n(N-1)}{N}S_b^2 - n\left(\frac{1}{n} - \frac{1}{N}\right)S_b^2 \\
&= (n-1)S_b^2 + (n-1)\left(\frac{1}{m} - \frac{1}{M}\right)\bar{S}_w^2.
\end{aligned} \tag{8.2.7}$$

Hence, we obtain

$$E(s_b^2) = S_b^2 + \left(\frac{1}{m} - \frac{1}{M}\right)\bar{S}_w^2. \tag{8.2.8}$$

Further,

$$\begin{aligned}
E(\bar{s}_w^2) &= E\left[\frac{1}{n}\sum_{i=1}^{n} s_i^2\right] \\
&= E_1\left[\frac{1}{n}\sum_{i=1}^{n} E_2(s_i^2|i)\right] \\
&= E_1\left[\frac{1}{n}\sum_{i=1}^{n} S_i^2\right] = \frac{1}{N}\sum_{i=1}^{N} S_i^2 = \bar{S}_w^2. \qquad (8.2.9)
\end{aligned}$$

Using (8.2.8) and (8.2.9), we get

$$E\left[s_b^2 - \left(\frac{1}{m} - \frac{1}{M}\right)\bar{s}_w^2\right] = S_b^2. \qquad (8.2.10)$$

Using (8.2.9) and (8.2.10), we obtain the unbiased estimator of the variance of $\bar{y}_{nm}$ as

$$\begin{aligned}
Est.Var(\bar{y}_{nm}) &= \left(\frac{1}{n} - \frac{1}{N}\right)\left[s_b^2 - \left(\frac{1}{m} - \frac{1}{M}\right)\bar{s}_w^2\right] + \frac{1}{n}\left(\frac{1}{m} - \frac{1}{M}\right)\bar{s}_w^2 \\
&= \left(\frac{1}{n} - \frac{1}{N}\right)s_b^2 + \frac{1}{N}\left(\frac{1}{m} - \frac{1}{M}\right)\bar{s}_w^2,
\end{aligned}$$

which proves the theorem.

Remark. Limiting cases:

(i) When N is large,

$$Est.Var(\bar{y}_{nm}) = \frac{s_b^2}{n}.$$

(ii) When M is large,

$$Est.Var(\bar{y}_{nm}) = \left(\frac{1}{n} - \frac{1}{N}\right)s_b^2 + \frac{\bar{s}_w^2}{Nm}.$$

(iii) When N and M both are large,

$$Est.Var(\bar{y}_{nm}) = \frac{s_b^2}{n}.$$

8.3 Comparison of Two-stage Sampling with One-stage Sampling

We shall compare the two-stage sampling with the following one-stage sampling procedures:

(i) SRSWOR of nm elements in one stage.
(ii) SRSWOR of $\frac{nm}{M}$ clusters (first-stage units) without further sub-sampling.

Theorem 8.3.1. *The efficiency of two-stage sampling relative to SRSWOR of nm elements in one stage sampling is given by*

$$E = \frac{1}{1+\rho\left(\frac{N-n}{N-1}m-1\right)},$$

assuming $\frac{m}{M}$ to be small.

Proof. We shall express the variance of $\bar{y}_{nm}$ in terms of ρ, the intraclass correlation coefficient between elements of first-stage units. From (7.2.5) and (7.2.6), we obtain

$$S_b^2 = \frac{(NM-1)S^2}{M^2(N-1)}[1+(M-1)\rho] \tag{8.3.1}$$

$$\bar{S}_w^2 = \frac{(NM-1)}{NM}s^2(1-\rho). \tag{8.3.2}$$

Substituting the values of S_b^2 and $\bar{S}_w^2$ from (8.3.1) and (8.3.2) in $Var(\bar{y}_{nm})$, after some simplification we obtain the variance of $\bar{y}_{nm}$ as

$$\begin{aligned} Var(\bar{y}_{nm})_{Two-st.} = & \frac{(NM-1)}{NM} \\ & \cdot\frac{S^2}{nm}\left[1-\frac{m(n-1)}{m(N-1)}\right. \\ & \left.+\rho\left\{\frac{(N-n)m}{(N-1)M}(M-1)-\frac{M-m}{M}\right\}\right]. \end{aligned} \tag{8.3.3}$$

Now, in single stage sampling of SRSWOR of nm elements, we have

$$Var(\bar{y}_{nm})_{SRS} = \left(\frac{1}{nm}-\frac{1}{NM}\right)S^2. \tag{8.3.4}$$

When sub-sampling rate $\frac{m}{M}$ is small, then (8.3.3) and (8.3.4) become

$$Var(\bar{y}_{nm})_{Two-st.} = \frac{S^2}{nm}\left[1+\rho\left(\frac{N-n}{N-1}m-1\right)\right] \tag{8.3.5}$$

$$Var(\bar{y}_{nm})_{SRS} = \frac{S^2}{nm}. \tag{8.3.6}$$

Hence, efficiency of two-stage relative to one-stage sampling of SRSWOR is obtained by dividing (8.3.6) by (8.3.5) and is given by

$$E = \frac{1}{1 + \rho\left(\frac{N-n}{N-1}m - 1\right)}. \tag{8.3.7}$$

Remark 1. If N is large, then (8.3.7) becomes

$$E = \frac{1}{1 + \rho(m-1)}. \tag{8.3.8}$$

Remark 2. We have seen in Chapter 7, that, when N is large, efficiency of cluster sampling is given by

$$E = \frac{1}{1 + \rho(M-1)}, \tag{8.3.9}$$

(refer to (7.3.6)). Hence, comparing (8.3.7) and (8.3.9), we find that the efficiency of two-stage sampling, when $\frac{m}{M}$ is small, is equal to that of cluster sampling in which the cluster size is $\frac{m(N-n)}{(N-1)}$.

We now compare the efficiency of two-stage sampling with that of a SRSWOR of $\frac{nm}{M}$ clusters surveyed completely. This is done in the following theorem.

Theorem 8.3.2. *Two-stage sampling is more, equally or less efficient than an equivalent SRSWOR of $\frac{nm}{M}$ clusters accordingly as*

$$S_b^2 - \frac{1}{M}\bar{S}_w^2 \gtreqless 0.$$

Proof. The sampling variance of a sample of $\frac{nm}{M}$ clusters is obtained from Theorem 7.2.1 as

$$Var(\bar{y}_c)_{\frac{nm}{M}} = \left(\frac{M}{nm} - \frac{1}{N}\right) S_b^2. \tag{8.3.10}$$

Subtracting (8.3.10) from (8.2.2), we get

$$Var(\bar{y}_{nm})_{Two-st.} - Var(\bar{y}_c)_{\frac{nm}{M}} = \frac{M}{n}\left(\frac{1}{m} - \frac{1}{M}\right)\left(S_b^2 - \frac{1}{M}\bar{S}_w^2\right). \tag{8.3.11}$$

Theorem follows from consideration of (8.3.11).

Remark. When N is large, from (8.3.1) and (8.3.2), we obtain

$$S_b^2 - \frac{1}{M}\bar{S}_w^2 = \rho S^2. \tag{8.3.12}$$

Hence, two-stage sampling is more efficient than equivalent cluster sampling of $\frac{nm}{M}$ clusters if $\rho > 0$, and smaller the sub-sampling rate $\frac{m}{M}$, the larger will be the reduction in variance of a two-stage sampling over equivalent cluster sampling.

8.4 Optimum Values of n and m

We see from Theorem 8.2.1 that the variance of a two-stage sampling , apart from the values of S_b^2 and $\bar{S}_w^2$, depends upon the values of n and m. The cost of the survey also depends upon n and m. In this section, we shall consider the problem of determination of values of n and m which will minimize the variance of the sample mean of a two-stage sample for a fixed cost of the survey. Alternatively, one can choose the values of n and m so that the cost of the survey is minimized for the given variance of the sample mean.

Theorem 8.4.1. *In two-stage sampling, assuming the cost function of the form $C = c_1 n + c_2 nm$, the optimum values of n and m, which will minimize the sampling variance for given cost C_0 of the survey are given by*

(i) *when* $S_b^2 - \frac{1}{M}\bar{S}_w^2 > 0$,

$$\hat{m} = \sqrt{\frac{c_1}{c_2} \cdot \frac{\bar{S}_w^2}{S_b^2 - \frac{1}{M}\bar{S}_w^2}}, \qquad \hat{n} = \frac{C_0}{c_1 + c_2\hat{m}}.$$

(ii) *when* $S_b^2 - \frac{1}{M}\bar{S}_w^2 \leq 0$,

$$\hat{m} = M, \quad \hat{n} = \frac{C_0}{c_1 + c_2 M}, \quad \text{if } C_0 \geq c_1 + c_2 M,$$

$$\hat{m} = \frac{C_0 - c_1}{c_2}, \quad \hat{n} = 1, \quad \text{if } C_0 < c_1 + c_2 M.$$

Proof. We are given the cost of the survey to be C_0. Hence, we have

$$C_0 = c_1 n + c_2 nm. \tag{8.4.1}$$

We minimize the sampling variance w.r.t. n and m subject to the restriction $C_0 = c_1 n + c_2 nm$. From the expession of the variance of $\bar{y}_{nm}$ and using

(8.4.1), we obtain

$$C_0[Var(\bar{y}_{nm}) + \frac{1}{N}S_b^2] = c_1\left(S_b^2 - \frac{1}{M}\bar{S}_w^2\right) + c_2\bar{S}_w^2$$
$$+ mc_2\left(S_b^2 - \frac{1}{M}\bar{S}_w^2\right) + \frac{1}{m}c_1\bar{S}_w^2. \tag{8.4.2}$$

When $S_b^2 - \frac{1}{M}\bar{S}_w^2 > 0$, (8.4.2) can be expressed as

$$C_0[Var(\bar{y}_{nm}) + \frac{1}{N}S_b^2] = \left[\sqrt{c_1\left(S_b^2 - \frac{1}{M}\bar{S}_w^2\right)} + \sqrt{c_2\bar{S}_w^2}\right]^2$$
$$+ \left[\sqrt{mc_2\left(S_b^2 - \frac{1}{M}\bar{S}_w^2\right)} - \sqrt{\frac{1}{m}c_1\bar{S}_w^2}\right]^2. \tag{8.4.3}$$

Clearly, (8.4.2) is minimum, when second term in (8.4.3) on r.h.s. is zero. Thus, optimum value $\hat{m}$ is obtained as

$$\hat{m} = \sqrt{\frac{c_1}{c_2} \cdot \frac{\bar{S}_w^2}{(S_b^2 - \frac{1}{M}\bar{S}_w^2)}}. \tag{8.4.4}$$

Then, optimum value of $\bar{n}$ follows from the relation $C_0 = c_1 n + c_2 nm$ and is obtained as

$$\hat{n} = \frac{C_0}{c_1 + c_2\hat{m}}. \tag{8.4.5}$$

Now, let us consider the case when $S_b^2 - \frac{1}{M}\bar{S}_w^2 \leq 0$. Then, from (8.4.1), it follows that (8.4.3) is minimum if m is the greatest attainable integer. Hence, in this case, when $C_0 \geq c_1 + c_2 M, ; \hat{m} = M$, and $\hat{n} = \frac{C_0}{(c_1+c_2M)}$. If however $C_0 < c_1 + c_2 M$, then $\hat{m} = \frac{C_0 - c_1}{c_2}$, and $\hat{n} = 1$.

Remark. When N is large, from (8.3.1) and (8.3.2), we obtain

$$\bar{S}_w^2 \cong S^2(1-\rho)$$
$$S_b^2 - \frac{1}{M}\bar{S}_w^2 \cong \rho S^2.$$

Hence, when N is large, $\hat{m}$ is given by

$$\hat{m} \cong \sqrt{\frac{c_1}{c_2}\left(\frac{1}{\rho} - 1\right)}. \tag{8.4.6}$$

8.5 Two-stage Sampling-unequal First-stage Units

We shall now develop the theory of two-stage sampling when the first-stage units are of unequal sizes. Suppose that the population is composed of N first-stage units and that the ith first-stage unit has M_i second-stage units, $i = 1, 2, \ldots, N$. Let

$$M_0 = \sum_{i=1}^{N} M_i = \text{total number of second-stage units in the population,}$$

and

$$\bar{M} = \frac{M_0}{N}.$$

Further, suppose that a SRSWOR of n first-stage units is selected and a SRSWOR of m_i second-stage units is selected from the selected i^{th} first-stage units, and

$$m_0 = \sum^{n} m_i = \text{the total number of second-stage units in the sample.}$$

Let

$$\bar{y}_{i\cdot} = \frac{1}{M_i}\sum_{j=1}^{M_i} y_{ij}$$

$$\bar{\bar{y}}_N = \frac{1}{N}\sum_{i=1}^{N} \bar{y}_{i\cdot}$$

$$\bar{y}_{\cdot\cdot} = \frac{1}{M_0}\sum_{i=1}^{N}\sum_{j=1}^{M_i} y_{ij} = \frac{1}{M_0}\sum_{i=1}^{N} M_i\bar{y}_{i\cdot} = \frac{1}{N}\sum_{i=1}^{N}\frac{M_i}{\bar{M}}\bar{y}_{i\cdot}$$

$$\bar{y}_{i(mi)} = \frac{1}{m_i}\sum_{j=1}^{m_i} y_{ij}$$

$$= \text{mean per second-stage unit in the } i^{th} \text{ selected first-stage in the sample.}$$

We first consider the estimation theory for the case when the values of M are known for the sample of first-stage units only.

Theorem 8.5.1.

(i) $\hat{Y} = \frac{N}{n}\sum_{i=1}^{n} M_i\bar{y}_{i(mi)}$ *is an unbiased estimator of the population total* $Y = \sum_{i=1}^{N}\sum_{j=1}^{M_i} y_{ij}$.

(ii) $Var(\hat{Y}) = N^2\bar{M}^2\left(\frac{1}{n} - \frac{1}{N}\right)S_b'^2 + \frac{N}{n}\sum_{i=1}^{N} M_i^2\left(\frac{1}{m_i} - \frac{1}{M_i}\right)S_i^2$, *where*

$$S_b'^2 = \frac{1}{N-1}\sum_{i=1}^{N}\left(\frac{M_i}{\bar{M}}\bar{y}_{i\cdot} - y_{\cdot\cdot}\right)^2, \text{and}$$

$$S_i^2 = \frac{1}{(M_i-1)}\sum_{j=1}^{M_i}(y_{ij} - \bar{y}_{i\cdot})^2.$$

Proof. We have

$$\begin{aligned}
E(\hat{Y}) &= E\left[\frac{N}{n}\sum_{i=1}^{n} M_i\bar{y}_{i(mi)}\right] \\
&= E_1\left[\frac{N}{n}\sum_{i=1}^{n} M_i E_2(\bar{y}_{i(mi)}|i)\right] \\
&= E_1\left[\frac{N}{n}\sum_{i=1}^{n} M_i\bar{y}_{i\cdot}\right] \\
&= N\cdot\frac{1}{N}\sum_{i=1}^{N} M_i\bar{y}_{i\cdot} \\
&= \sum_{i=1}^{N} M_i\bar{y}_{i\cdot} = Y. \qquad (8.5.1)
\end{aligned}$$

Next, consider the variance of $\hat{Y}$.

$$\begin{aligned}
Var(\hat{Y}) &= Var\left(\frac{N}{n}\sum_{i=1}^{n} M_i\bar{y}_{i(mi)}\right) \\
&= V_1\left[\frac{N}{n}\sum_{i=1}^{n} M_i E_2(\bar{y}_{i(mi)}|i)\right] + E_2V_1\left[\frac{N}{n}\sum_{i=1}^{n} M_i(\bar{y}_{i(mi)}|i)\right]
\end{aligned}$$

$$= V_1\left[\tfrac{N}{n}\sum_{i=1}^{n} M_i\bar{y}_{i\cdot}\right] + \tfrac{N^2}{n^2}E_2\left[\sum_{i=1}^{n} M_i^2\left(\tfrac{1}{n_i}-\tfrac{1}{M_i}\right)S_i^2\right]$$

$$= N^2\cdot\left(\frac{1}{n}-\frac{1}{N}\right)\frac{\sum_{i=1}^{N}(M_i\bar{y}_{i\cdot}-\bar{M}\bar{y}_{\cdot\cdot})^2}{N-1}$$

$$+\frac{N^2}{n}\cdot\frac{1}{N}\sum_{i=1}^{N} M_i^2\left(\frac{1}{m_i}-\frac{1}{M_i}\right)S_i^2$$

$$= N^2\bar{M}^2\cdot\left(\frac{1}{n}-\frac{1}{N}\right)S_b'^2 + \frac{N}{n}\sum_{i=1}^{N} M_i^2\left(\frac{1}{n_i}-\frac{1}{M_i}\right)S_i^2,$$

which proves the theorem.

Theorem 8.5.2. *An unbiased estimator of variance of* $\hat{Y} = \frac{N}{n}\sum_{i=1}^{n} M_i\bar{y}_{i(mi)}$ *is given by*

$$Est.Var(\hat{Y}) = N^2\bar{M}^2\left(\frac{1}{n}-\frac{1}{N}\right)s_b'^2 + \frac{N}{n}\sum_{i=1}^{n} M_i^2\left(\frac{1}{m_i}-\frac{1}{M_i}\right)s_i^2$$

where

$$s_b'^2 = \frac{1}{(n-1)}\sum_{i=1}^{n}\left(\frac{M_i}{\bar{M}}\bar{y}_{i(mi)}-\hat{\bar{Y}}\right)^2,$$

$$\hat{\bar{Y}} = \frac{1}{n}\sum_{i=1}^{n}\frac{M_i}{\bar{M}}\bar{y}_{i(mi)} = \frac{\hat{Y}}{N\bar{M}}$$

$$s_i^2 = \frac{1}{(m_i-1)}\sum_{j=1}^{m_i}(y_{ij}-\bar{y}_{i(mi)})^2.$$

Proof. We note that

$$(n-1)\bar{M}^2 s_b'^2 = \sum_{i=1}^{n} M_i^2\bar{y}_{i(mi)}^2 - \frac{n}{N^2}\hat{Y}^2.$$

Hence,

$$\begin{aligned}
(n-1)\bar{M}^2 E(s_b'^2) &= E_1\left[\sum_{i=1}^{n} M_i^2 E_2(\bar{y}_{i(mi)}^2|i)\right] - \frac{n}{N^2}[Var(\hat{Y}) + Y^2] \\
&= E_1\left[\sum_{i=1}^{n} M_i^2\left(\frac{1}{m_i} - \frac{1}{M_i}\right) S_i^2 + \sum_{i=1}^{n} M_i^2 \bar{y}_{i\cdot}^2\right] \\
&\quad - \frac{n}{N^2} Var(\hat{Y}) - \frac{n}{N^2} Y^2 \\
&= \frac{n}{N}\sum_{i=1}^{N} M_i^2\left(\frac{1}{m_i} - \frac{1}{M_i}\right) S_i^2 + \frac{n}{N}\sum_{i=1}^{N} M_i^2 \bar{y}_{i\cdot}^2 \\
&\quad - \frac{n}{N^2} Y^2 - \frac{n}{N^2} Var(\hat{Y}) \\
&= \frac{n}{N}\sum_{i=1}^{n} M_i^2\left(\frac{1}{m_i} - \frac{1}{M_i}\right) S_i^2 + \frac{n\bar{M}^2(N-1)}{N} S_b'^2 \\
&\quad - \frac{n}{N^2} Var(\hat{Y}) \\
&= \frac{n}{N}\left[\sum_{i=1}^{N} M_i^2\left(\frac{1}{m_i} - \frac{1}{M_i}\right) S_i^2 - \frac{1}{N} Var(\hat{Y})\right. \\
&\quad \left. + \frac{n\bar{M}^2(N-1)}{N} S_b'^2\right] \\
&= \frac{n}{N}\left[\sum_{i=1}^{N} M_i^2\left(\frac{1}{m_i} - \frac{1}{M_i}\right) S_i^2 - N\bar{M}^2\left(\frac{1}{n} - \frac{1}{N}\right) S_b'^2\right. \\
&\quad \left. - \frac{1}{n}\sum_{i=1}^{N} M_i^2\left(\frac{1}{m_i} - \frac{1}{M_i}\right) S_i^2\right] + \frac{n\bar{M}^2(N-1)}{N} S_b'^2 \\
&= \frac{(n-1)}{N}\sum_{i=1}^{N} M_i^2\left(\frac{1}{m_i} - \frac{1}{M_i}\right) S_i^2 - n\bar{M}^2\left(\frac{1}{n} - \frac{1}{N}\right) S_b'^2 \\
&\quad + \frac{n\bar{M}^2(N-1)}{N} S_b'^2 \\
&= \frac{(n-1)}{N}\sum_{i=1}^{N} M_i^2\left(\frac{1}{m_i} - \frac{1}{M_i}\right) S_i^2 + (n-1)\bar{M}^2 S_b'^2.
\end{aligned}$$

Hence

$$E(s_b'^2) = \frac{1}{N\bar{M}^2}\sum_{i=1}^{N} M_i^2\left(\frac{1}{m_i} - \frac{1}{M_i}\right) S_i^2 + S_b'^2. \tag{8.5.2}$$

Further, since $E(s_i^2) = S_i^2$, we have

$$E\left[s_b'^2 - \frac{1}{n\bar{M}^2}\sum_{i=1}^{n} M_i^2\left(\frac{1}{m_i} - \frac{1}{M_i}\right)s_i^2\right] = S_b'^2. \tag{8.5.3}$$

Hence, an unbiased estimator of $Var(\hat{Y})$ is given by

$$\begin{aligned} Est.Var(\hat{Y}) &= N^2\bar{M}^2\left(\frac{1}{n} - \frac{1}{N}\right)\left[s_b'^2 - \frac{1}{n\bar{M}^2}\sum_{i=1}^{n} M_i^2\left(\frac{1}{m_i} - \frac{1}{M_i}\right)s_i^2\right] \\ &\quad + \frac{N}{n}\cdot\frac{N}{n}\sum_{i=1}^{n} M_i^2\left(\frac{1}{m_i} - \frac{1}{M_i}\right)s_i^2 \\ &= N^2\bar{M}^2\left(\frac{1}{n} - \frac{1}{N}\right)s_b'^2 + \frac{N}{n}\sum_{i=1}^{n} M_i^2\left(\frac{1}{m_i} - \frac{1}{M_i}\right)s_i^2. \end{aligned} \tag{8.5.4}$$

It may be noted that for calculating the first term in (8.5.4), the value of M is not required, since it can be written as

$$N^2\bar{M}^2\left(\frac{1}{n} - \frac{1}{N}\right)s_b'^2 = \frac{(N-n)}{Nn(n-1)}\sum_{i=1}^{n}(NM_i\bar{y}_{i(mi)} - \hat{Y})^2.$$

We may note here that although the population total Y is unbiasedly estimated by $\hat{Y} = \frac{N}{n}\sum_{i=1}^{n} M_i\bar{y}_{i\cdot}$, the population mean is not estimated by $\frac{\hat{Y}}{N\bar{M}}$, since $\bar{M}$ is not known.

We now consider estimators of the population mean $\bar{y}_{\cdot\cdot}$ when the values of M_i are known only for the first-stage units of the sample. These are given by

(i) $\bar{\bar{y}}'_n = \frac{1}{n}\sum_{i=1}^{n}\bar{y}_{i(mi)}$ = the mean of the first-stage unit means in the sample.

(ii) $\bar{y}'_{nR} = \dfrac{\sum_{i=1}^{n} M_i\bar{y}_{i(mi)}}{\sum_{i=1}^{n} M_i}$.

(i) $\bar{\bar{y}}'_{n\cdot}$: Bias and Mean Square Error

We have

$$E[\bar{\bar{y}}'_n] = E\left[\frac{1}{n}\sum_{i=1}^{n}\bar{y}_{i(mi)}\right]$$
$$= E_1\left[\frac{1}{n}\sum_{i=1}^{n}E_2(\bar{y}_{i(mi)}|i)\right]$$
$$= E_1\left[\frac{1}{n}\sum_{i=1}^{n}\bar{y}_{i\cdot}\right]$$
$$= \frac{1}{N}\sum_{i=1}^{N}\bar{y}_{i\cdot} = \bar{\bar{y}}_N \neq \bar{y}_{\cdot\cdot}\,. \tag{8.5.5}$$

Thus $\bar{\bar{y}}'_n$ is a biased estimator of $\bar{y}_{\cdot\cdot}$, and

$$Bias(\bar{\bar{y}}'_n) = \bar{\bar{y}}_N - \bar{y}_{\cdot\cdot}. \tag{8.5.6}$$

Further,

$$Var(\bar{\bar{y}}'_n) = V_1E_2\left(\frac{1}{n}\sum_{i=1}^{n}\bar{y}_{i(mi)}\right) + E_1V_2\left(\frac{1}{n}\sum_{i=1}^{n}\bar{y}_{i(mi)}\right)$$
$$= V_1\left[\frac{1}{n}\sum_{i=1}^{n}E_2(\bar{y}_{i(mi)}|i)\right] + E_1\left[\frac{1}{n}\sum_{i=1}^{n}V_2(\bar{y}_{i(mi)}|i)\right]$$
$$= V_1\left[\frac{1}{n}\sum_{i=1}^{n}\bar{y}_{i\cdot}\right] + \frac{1}{n^2}E_1\left[\sum_{i=1}^{n}\left(\frac{1}{m_i}-\frac{1}{M_i}\right)S_i^2\right]$$
$$= \left(\frac{1}{n}-\frac{1}{N}\right)\cdot S_b^2 + \frac{1}{nN}\sum_{i=1}^{N}\left(\frac{1}{m_i}-\frac{1}{M_i}\right)S_i^2, \tag{8.5.7}$$

where

$$S_b^2 = \frac{1}{N-1}\sum_{i=1}^{N}(\bar{y}_{i\cdot}-\bar{\bar{y}}_N)^2, \quad \bar{\bar{y}}_N = \frac{1}{N}\sum_{i=1}^{N}\bar{y}_{i\cdot}$$

$$S_i^2 = \frac{1}{M_i-1}\sum_{j=1}^{M_i}(y_{ij}-\bar{y}_{i\cdot})^2.$$

Hence, the mean square error of $\bar{\bar{y}}'_n$ is given by

$$MSE(\bar{\bar{y}}'_n) = Var(\bar{\bar{y}}'_n) + [Bias]^2$$
$$= \left(\frac{1}{n}-\frac{1}{N}\right)S_b^2 + \frac{1}{nN}\sum_{i=1}^{N}\left(\frac{1}{m_i}-\frac{1}{M_i}\right)S_i^2 + (\bar{\bar{y}}_N - \bar{y}_{\cdot\cdot})^2. \tag{8.5.8}$$

Remark 1. The bias of $\bar{\bar{y}}'_n$ can be written as

$$\begin{aligned}
Bias(\bar{\bar{y}}'_n) &= \bar{\bar{y}}_N - \bar{y}_{..} \\
&= \frac{1}{N}\sum_{i=1}^{N}\bar{y}_{i\cdot} - \frac{1}{N\bar{M}}\sum_{i=1}^{N}M_i\bar{y}_{i\cdot} \\
&= \frac{1}{N\bar{M}}\left[\sum_{i=1}^{N}M_i\bar{y}_{i\cdot} - \frac{1}{N}\left(\sum_{i=1}^{N}\bar{y}_{i\cdot}\right)\left(\sum_{i=1}^{N}M_i\right)\right] \\
&= -\frac{1}{N\bar{M}}\sum_{i=1}^{N}(M_i\bar{M})(\bar{y}_{i\cdot} - \bar{\bar{y}}_N). \qquad (8.5.9)
\end{aligned}$$

Hence, an unbiased estimator of the bias is obtained as

$$Est.Bias(y') = -\frac{N-1}{N\bar{M}} \cdot \frac{\sum_{i=1}^{n}(M_i - \bar{M}_n)(y_{i(mi)} - \bar{\bar{y}}'_n)}{(n-1)} \qquad (8.5.10)$$

where $\bar{M}_n = \frac{1}{n}\sum_{i=1}^{n} M_i$.

Remark 2. Using Remark 1, we obtain an unbiased estimator of $\bar{y}_{..}$ as

$$\bar{\bar{y}}'_n + \frac{N-1}{N\bar{M}} \cdot \frac{\sum_{i=1}^{n}(M_i - \bar{M}_n)(\bar{y}_{i(mi)} - \bar{\bar{y}}'_n)}{(n-1)}. \qquad (8.5.11)$$

Remark 3. An unbiased estimator of the variance of $\bar{\bar{y}}'_n$ is given by

$$Est.Var(\bar{\bar{y}}'_n) = \left(\frac{1}{n} - \frac{1}{N}\right)s_b^2 + \frac{1}{nN}\sum_{i=1}^{n}\left(\frac{1}{m_i} - \frac{1}{M_i}\right)s_i^2, \qquad (8.5.12)$$

where

$$s_b^2 = \frac{1}{(n-1)}\sum_{i=1}^{n}(\bar{y}_{i(mi)} - \bar{\bar{y}}'_n)^2.$$

Proof.

$$(n-1)s_b^2 = \sum_{i=1}^{n}\bar{y}_{i(mi)}^2 - n\bar{\bar{y}}'_n.$$

Hence,

$$
\begin{aligned}
(n-1)E(s_b^2) &= E_1\left[\sum_{i=1}^{n} E_2(\bar{y}_{i(mi)}^2|i)\right] - n\,Var(\bar{\bar{y}}_n') - n\bar{\bar{y}}_N^2 \\
&= E_1\left[\sum_{i=1}^{n}\left(\frac{1}{m_i}-\frac{1}{M_i}\right)S_i^2 + \sum_{i=1}^{n}\bar{y}_{i.}^2\right] \\
&\quad - n\bar{\bar{y}}_N^2 - n\,Var(\bar{\bar{y}}_n') \\
&= \frac{n}{N}\sum_{i=1}^{N}\left(\frac{1}{m_i}-\frac{1}{M_i}\right)S_i^2 + \frac{n}{N}\sum_{i=1}^{N}\bar{y}_{i.}^2 \\
&\quad - n\bar{\bar{y}}_N^2 - n\,Var(\bar{\bar{y}}_n') \\
&= (n-1)S_b^2 + \frac{(n-1)}{N}\sum_{i=1}^{N}\left(\frac{1}{m_i}-\frac{1}{M_i}\right)S_i^2,
\end{aligned}
$$

which gives

$$
E(s_b^2) = S_b^2 + \frac{1}{N}\sum_{i=1}^{N}\left(\frac{1}{m_i}-\frac{1}{M_i}\right)S_i^2. \tag{8.5.13}
$$

Now

$$
\begin{aligned}
E\left[\frac{1}{n}\sum_{i=1}^{n}\left(\frac{1}{m_i}-\frac{1}{M_i}\right)s_i^2\right] &= E_1\left[\frac{1}{n}\sum_{i=1}^{n}\left(\frac{1}{m_i}-\frac{1}{M_i}\right)E_2(s_i^2|i)\right] \\
&= E_1\left[\frac{1}{n}\sum_{i=1}^{n}\left(\frac{1}{m_i}-\frac{1}{M_i}\right)S_i^2\right] \\
&= \frac{1}{N}\left[\sum_{i=1}^{N}\left(\frac{1}{m_i}-\frac{1}{M_i}\right)S_i^2\right].
\end{aligned} \tag{8.5.14}
$$

From (8.5.13) and (8.5.14), one obtains

$$
E\left[s_b^2 - \frac{1}{n}\sum_{i=1}^{n}\left(\frac{1}{m_i}-\frac{1}{M_i}\right)s_i^2\right] = S_b^2.
$$

Hence, an unbiased estimator of the variance is given by

$$
Est.Var(\bar{\bar{y}}_n') = \left(\frac{1}{n}-\frac{1}{N}\right)s_b^2 + \frac{1}{nN}\sum_{i=1}^{n}\left(\frac{1}{m_i}-\frac{1}{M_i}\right)s_i^2. \tag{8.5.15}
$$

Remark 4. An estimator of the $MSE(\bar{\bar{y}}'_n)$ is given by

$$Est.MSE(\bar{\bar{y}}'_n) = \left(\frac{1}{n} - \frac{1}{N}\right) s_b^2 + \frac{1}{nN}\sum_{i=1}^{n}\left(\frac{1}{m_i} - \frac{1}{M_i}\right) s_i^2 + \frac{(N-1)^2}{N^2\bar{M}^2}\left[\sum_{i=1}^{n}(M_i - \bar{M}_n)(\bar{y}_{i(mi)} - \bar{\bar{y}}'_n)^2\right].$$

(ii) $\bar{y}_{nR'}$:

The estimator $\bar{y}_{nR'}$ can be written as

$$\bar{y}_{nR'} = \frac{\sum_{i=1}^{n} M_i \bar{y}_{i(mi)}}{\sum_{i=1}^{n} M_i} = \frac{\frac{1}{n}\sum_{i=1}^{n} u_i \bar{y}_{i(mi)}}{\frac{1}{n}\sum_{i=1}^{n} u_i}$$

where $u_i = \frac{M_i}{\bar{M}}$, and $\bar{M} = \frac{\sum_{i=1}^{N} M_i}{N}$.

We consider a more general ratio estimator by defining x_{ij} as the value of an auxiliary variable x corresponding to the value y_{ij} of y, the variable under study. Let

$$\bar{x}_{..} = \frac{\sum_{i=1}^{N}\sum_{j=1}^{M_i} x_{ij}}{N\bar{M}} = \frac{1}{N}\sum_{i=1}^{N} u_i \bar{x}_{i.},$$

$$\bar{x}_{i.} = \frac{1}{M_i}\sum_{j=1}^{M_i} x_{ij}$$

$$\bar{x}_{i(mi)} = \frac{1}{m_i}\sum_{j=1}^{m_i} x_{ij}.$$

We consider a general ratio estimator of $\bar{y}_{..}$ as

$$\bar{y}_{GR'} = \frac{\frac{1}{n}\sum_{i=1}^{n} y_i \bar{y}_{i(mi)}}{\frac{1}{n}\sum_{i=1}^{n} u_i \bar{x}_{i(mi)}}\bar{x}_{..}\,. \tag{8.5.17}$$

If $x_{ij} = 1$ for all i and j, then $\bar{y}_{GR'} = \bar{y}_{nR'}$. Denote $\frac{1}{n}\sum_{i=1}^{n} u_i\bar{y}_{i(mi)}$ and $\frac{1}{n}\sum_{i=1}^{n} u_i\bar{x}_{i(mi)}$ by $\bar{y}'_w$ and $\bar{x}_w$ respectively. Note that

$$E(\bar{y}_w) = E\left(\frac{1}{n}\sum_{i=1}^{n} u_i\bar{y}_{i(mi)}\right)$$
$$= E_1\left[\frac{1}{n}\sum_{i=1}^{n} u_i E_2(\bar{y}_{i(mi)}|i)\right]$$
$$= E_1\left[\frac{1}{n}\sum_{i=1}^{n} u_i\bar{y}_{i\cdot}\right]$$
$$= \frac{1}{N}\sum_{i=1}^{n} u_i\bar{y}_{i\cdot} = \bar{y}_{\cdot\cdot}\,. \tag{8.5.18}$$

Similarly, we have

$$E(\bar{x}_w) = \bar{x}_{\cdot\cdot}\,. \tag{8.5.19}$$

We apply the theory of ratio method of estimation developed in Chapter 5 to investigate the properties of $\bar{y}_{GR'}$. Using Theorem 5.2.1, we obtain the bias of $\bar{y}_{GR'}$ as

$$Bias(\bar{y}_{GR'}) = \frac{1}{\bar{x}_{\cdot\cdot}}[RVar(\bar{x}_w) - Cov(\bar{y}_w, \bar{x}_w)]. \tag{8.5.20}$$

Also, from (5.3.2), it follows that the variance of $\bar{y}_{GR'}$ to the first order of approximation is given by

$$Var(\bar{y}_{GR'}) = Var(\bar{y}_w) + R^2Var(\bar{x}_w) - 2RCov(\bar{y}_w, \bar{x}_w), \tag{8.5.21}$$

where $R = \frac{\bar{y}_{\cdot\cdot}}{\bar{x}_{\cdot\cdot}}$.

We must now evaluate $Var(\bar{y}_w)$, $Var(\bar{x}_w)$ and $Cov(\bar{y}_w, \bar{x}_w)$. This is done as follows:

$$Var(\bar{y}_w) = Var\left(\frac{1}{n}\sum_{i=1}^{n} u_i\bar{y}_{i(mi)}\right)$$
$$= V_1\left[\frac{1}{n}\sum_{i=1}^{n} u_i E_2(\bar{y}_{i(mi)}|i)\right] + E_1V_2\left[\frac{1}{n}\sum_{i=1}^{n} u_i\bar{y}_{i(mi)}|i\right]$$
$$= V_1\left[\frac{1}{n}\sum_{i=1}^{n} u_i\bar{y}_{i\cdot}\right) + \frac{1}{n}E_1\left[\frac{1}{n}\sum_{i=1}^{n} u_i^2\left(\frac{1}{m_i} - \frac{1}{M_i}\right)S_{iy}^2\right]$$
$$= \left(\frac{1}{n} - \frac{1}{N}\right)S_{by}'^2 + \frac{1}{n}\cdot\frac{1}{N}\sum_{i=1}^{N} u_i^2\left(\frac{1}{m_i} - \frac{1}{M_i}\right)S_{iy}^2, \tag{8.5.22}$$

where

$$S'^2_{by} = \frac{1}{(N-1)} \sum_{i=1}^{N} (u_i \bar{y}_{i\cdot} - \bar{y}_{\cdot\cdot})^2$$

$$S^2_{iy} = \frac{1}{(M_i - 1)} \sum_{j=1}^{M_i} (y_{ij} - \bar{y}_{i\cdot})^2.$$

Replacing y_{ij} by x_{ij} in (8.5.22), we obtain $Var(\bar{x}_w)$ as

$$Var(\bar{x}_w) = \left(\frac{1}{n} - \frac{1}{N}\right) S'^2_{bx} + \frac{1}{nN} \sum_{i=1}^{N} u_i^2 \left(\frac{1}{m_i} - \frac{1}{M_i}\right) S^2_{ix}, \qquad (8.5.23)$$

where

$$S'^2_{bx} = \frac{1}{(N-1)} \sum_{i=1}^{N} (u_i \bar{x}_{i\cdot} - \bar{x}_{\cdot\cdot})^2$$

$$S^2_{ix} = \frac{1}{(M_i - 1)} \sum_{j=1}^{M_i} (x_{ij} - \bar{x}_{i\cdot})^2.$$

Let us now find $Cov(\bar{y}_w, \bar{x}_w)$.

$$\begin{aligned}
Cov(\bar{y}_w, \bar{x}_w) &= Cov\left(\frac{1}{n}\sum_{i=1}^{n} u_i \bar{y}_{i(mi)},\ \frac{1}{n}\sum_{i=1}^{n} u_i \bar{x}_{i(mi)}\right) \\
&= Cov\left[\frac{1}{n}\sum_{i=1}^{n} u_i E(\bar{y}_{i(mi)}|i),\ \frac{1}{n}\sum_{i=1}^{n} u_i E(\bar{x}_{i(mi)}|i)\right] \\
&\quad + E\left[\frac{1}{n^2}\sum_{i=1}^{n} u_i^2 Cov(\bar{y}_{i(mi)},\ \bar{x}_{i(mi)}|i)\right] \\
&= Cov\left(\frac{1}{n}\sum_{i=1}^{n} u_i \bar{y}_{i\cdot},\ \frac{1}{n}\sum_{i=1}^{n} u_i \bar{x}_{i\cdot}\right) \\
&\quad + \frac{1}{n} \cdot E\left[\frac{1}{n}\sum_{i=1}^{n} u_i^2 \left(\frac{1}{m_i} - \frac{1}{M_i}\right) S_{ixy}\right] \\
&= \left(\frac{1}{n} - \frac{1}{N}\right) S'_{bxy} + \frac{1}{nN}\sum_{i=1}^{N} u_i^2 \left(\frac{1}{m_i} - \frac{1}{M_i}\right) S_{ixy}, \qquad (8.5.24)
\end{aligned}$$

where

$$S'_{bxy} = \frac{1}{N-1} \sum_{i=1}^{N} (u_i \bar{x}_{i\cdot} - \bar{x}_{\cdot\cdot})(u_i \bar{y}_{i\cdot} - \bar{y}_{\cdot\cdot})$$

$$S'_{ixy} = \frac{1}{M_i - 1} \sum_{j=1}^{M_i} (x_{ij} - \bar{x}_{i\cdot})(y_{ij} - \bar{y}_{i\cdot}).$$

Substituting of (8.5.22), (8.5.23), and (8.5.24) in (8.5.20) and (8.5.21), we obtain

$$Bias(\bar{y}_{GR'}) = \frac{1}{\bar{x}_{..}}\left[\left(\frac{1}{n}-\frac{1}{N}\right)(RS'^2_{bx} - S'_{bxy}) + \frac{1}{nN}\sum_{i=1}^{n} u_i^2\left(\frac{1}{m_i}-\frac{1}{M_i}\right)(rS^2_{ix} - S_{ixy})\right] \quad (8.5.25)$$

$$Var(\bar{y}_{GR'}) = \left(\frac{1}{n}-\frac{1}{N}\right)D^2 + \frac{1}{nN}\sum_{i=1}^{N} u_i^2\left(\frac{1}{m_i}-\frac{1}{M_i}\right)D_i^2, \quad (8.5.26)$$

where

$$D^2 = S'^2_{by} + R^2 S'^2_{bx} - 2RS'_{bxy}$$
$$D_i^2 = S^2_{iy} + R^2 S^2_{ix} - 2RS_{ixy}.$$

Note that D^2 can also be written as

$$D^2 = \frac{1}{(N-1)}\sum_{i=1}^{N} u_i^2(\bar{y}_{i\cdot} - R\bar{x}_{i\cdot})^2. \quad (8.5.27)$$

Taking $x_{ij} = 1$ for all i and j, we find that $\bar{y}_{GR'}$ becomes $\bar{y}_{nR'}$. Hence, the bias and mean square error of $\bar{y}_{nR'}$ are obtained from (8.5.25) and (8.5.26) by taking $x_{ij} = 1$ for i and j and are given by

$$Bias(\bar{y}_{nR'}) = \left(\frac{1}{n}-\frac{1}{N}\right)(\bar{y}_{..}S_u^2 - S'_{uy}) \quad (8.5.28)$$

$$MSE(\bar{y}_{nR'}) = \left(\frac{1}{n}-\frac{1}{N}\right)S''^2_b + \frac{1}{nN}\sum_{i=1}^{N} u_i^2\left(\frac{1}{m_i}-\frac{1}{M_i}\right)S_i^2, \quad (8.5.29)$$

where

$$S''^2_b = \frac{1}{N-1}\sum_{i=1}^{N} u_i^2(\bar{y}_{i\cdot} - \bar{y}_{..})^2$$
$$S_i^2 = \frac{1}{M_i-1}\sum_{j=1}^{M_i}(y_{ij} - \bar{y}_{i\cdot})^2.$$

8.6 Three-stage Sampling

We now consider the theory of three-stage sampling. The population is assumed to be composed of N first-stage units and each first-stage unit has M second-stage units, and each second-stage unit has P third-stage units. A sample of n first-stage units is selected with SRSWOR and from each selected first-stage unit, a sample of m second-stage units is selected with SRSWOR, and from each selected second-stage, a further sample of p third-stage units is selected with SRSWOR. Note that the population consists of NMP elements and the sample consists of nmp elements. The following notations are followed.

y_{ijk} = the value of kth third-stage unit in the jth second-stage unit of the ith first-stage unit $k = 1, 2, \ldots, P$; $j = 1, 2, \ldots, M$; $i = 1, 2, \ldots, N$.

$$\bar{y}_{ij\cdot} = \frac{1}{P}\sum_{k=1}^{P} y_{ijk}$$

$$\bar{y}_{i\cdot\cdot} = \frac{1}{MP}\sum_{j=1}^{M}\sum_{k=1}^{P} y_{ijk} = \frac{1}{M}\sum_{j=1}^{M}\bar{y}_{ij\cdot}$$

$$\begin{aligned}\bar{\bar{y}}_{\cdots} &= \frac{1}{NMP}\sum_{i=1}^{N}\sum_{j=1}^{M}\sum_{k=1}^{P} y_{ijk} \\ &= \frac{1}{NM}\sum_{i=1}^{N}\sum_{j=1}^{M}\bar{y}_{ij\cdot} \\ &= \frac{1}{N}\sum_{i=1}^{N}\bar{y}_{i\cdot\cdot}\,.\end{aligned}$$

For the sample, we define the corresponding quantities $\bar{y}_{ij(p)}$, $\bar{y}_{i(mp)}$, $\bar{y}_{(nmp)}$ similarly.

Theorem 8.6.1.

(i) $\bar{y}_{(nmp)}$ *is an unbiased estimator of the population mean.*

(ii) $Var(\bar{y}_{(nmp)}) = \left(\frac{1}{n} - \frac{1}{N}\right) S_b^2 + \frac{1}{n}\left(\frac{1}{m} - \frac{1}{M}\right) \bar{S}_w^2 + \frac{1}{nm}\left(\frac{1}{p} - \frac{1}{P}\right) \bar{\bar{S}}_p^2.$

Proof. We have

$$E(\bar{y}_{(nmp)}) = E_1E_2E_3\left[\frac{1}{nmp}\sum_{i=1}^{n}\sum_{j=1}^{n}\sum_{k=1}^{p}y_{ijk}\right]$$

$$= E_1E2\left[\frac{1}{nm}\sum_{i=1}^{n}\sum_{j=1}^{m}E_3(\bar{y}_{ij(p)}|i,j)\right]$$

$$= E_1E_2\left[\frac{1}{nm}\sum_{i=1}^{n}\sum_{j=1}^{m}\bar{y}_{ij.}\right]$$

$$= E_1\left[\frac{1}{n}\sum_{i=1}^{n}E_2\left(\frac{1}{m}\sum_{j=1}^{m}\bar{y}_{ij.}|i\right)\right]$$

$$= E_1\left[\frac{1}{n}\sum_{i=1}^{n}\bar{y}_{i..}\right] = \frac{1}{N}\sum_{i=1}^{N}\bar{y}_{i..} = \bar{y}_{...} \tag{8.6.1}$$

Now,

$$Var(\bar{y}_{(nmp)}) = Var\left(\frac{1}{n}\sum_{i=1}^{n}\bar{y}_{i(mp)}\right)$$

$$= V_1E_2\left(\frac{1}{n}\sum_{i=1}^{n}\bar{y}_{i(mp)}|i\right) + E_1V_2\left(\frac{1}{n}\sum_{i=1}^{n}\bar{y}_{i(mp)}|i\right)$$

$$= V_1\left[\frac{1}{n}\sum_{i=1}^{n}E_2(\bar{y}_{i(mp)}|i)\right] + E_1\left[\frac{1}{n^2}\sum_{i=1}^{n}V_2(\bar{y}_{i(mp)}|i)\right]$$

$$= V_1\left[\frac{1}{n}\sum_{i=1}^{n}\bar{y}_{i..}\right]$$

$$+\frac{1}{n^2}E_1\left[\sum_{i=1}^{n}\left\{\left(\frac{1}{m}-\frac{1}{M}\right)S_i^2 + \frac{1}{m}\left(\frac{1}{p}-\frac{1}{P}\right)\frac{\sum_{j=1}^{M}S_{ij}^2}{M}\right\}\right] \tag{8.6.2}$$

where

$$S_i^2 = \frac{1}{M-1}\sum_{j=1}^{M}(\bar{y}_{ij\cdot} - \bar{y}_{i\cdot\cdot})^2$$

$$S_{ij}^2 = \frac{1}{P-1}\sum_{k=1}^{P}(y_{ijk} - \bar{y}_{ij\cdot})^2.$$

Then, considering (8.6.2), we get

$$\begin{aligned} Var(\bar{y}_{(nmp)}) &= \left(\frac{1}{n} - \frac{1}{N}\right) S_b^2 + \frac{1}{nN}\sum_{i=1}^{N}\left(\frac{1}{m} - \frac{1}{M}\right) S_i^2 \\ &\quad + \frac{1}{nN}\sum_{i=1}^{N}\frac{1}{m}\left(\frac{1}{p} - \frac{1}{P}\right)\frac{\sum_{j=1}^{M} S_{ij}^2}{M} \\ &= \left(\frac{1}{n} - \frac{1}{N}\right) S_b^2 + \frac{1}{n}\left(\frac{1}{m} - \frac{1}{M}\right)\bar{S}_w^2 + \frac{1}{nm}\left(\frac{1}{p} - \frac{1}{P}\right)\bar{\bar{S}}_p^2, \end{aligned} \tag{8.6.3}$$

where

$$S_b^2 = \frac{1}{N-1}\sum_{i=1}^{N}(\bar{y}_{i\cdot\cdot} - \bar{y}_{\cdot\cdot\cdot})^2$$

$$\bar{S}_w^2 = \frac{1}{N}\sum_{i=1}^{N} S_i^2$$

$$\bar{\bar{S}}_p^2 = \frac{1}{NM}\sum_{i=1}^{N}\sum_{j=1}^{M} S_{ij}^2.$$

Remark. We note that the variance of the sample mean consists of three components. If each of the nm selected second-stage units were completely enumerated, that is, $p = P$, then the variance of the sample mean would be given by the first two terms appropriate to two-stage sampling. If each unit were completely enumerated, then the variance would be given by the first term appropriate to cluster sampling.

When $n = N$, then the variance is given by the second and the third terms. This case corresponds to a stratified two-stage sampling, with first-stage units in the population forming the strata.

If N, M and P are large, then the variance of $\bar{y}_{(nmp)}$ is given by

$$Var(\bar{y}_{(nmp)}) = \frac{S_b^2}{n} + \frac{\bar{S}_w^2}{nm} + \frac{\bar{\bar{S}}_p^2}{nmp}. \tag{8.6.4}$$

In the following theorem, we give an unbiased estimator of the variance of $\bar{y}_{(nmp)}$.

Theorem 8.6.2. *An unbiased estimator of the variance of $\bar{y}_{(nmp)}$ is given by*

$$Est.Var(y_{(nmp)}) = \left(\frac{1}{n} - \frac{1}{N}\right) s_b^2 + \frac{1}{N}\left(\frac{1}{m} - \frac{1}{M}\right) \bar{s}_w^2 + \frac{1}{NM}\left(\frac{1}{p} - \frac{1}{P}\right) \bar{\bar{s}}_p^2,$$

where

$$s_b^2 = \frac{1}{(n-1)} \sum_{i=1}^{n} (\bar{y}_{i(mp)} - \bar{y}_{(nmp)})^2$$

$$s_i^2 = \frac{1}{(m-1)} \sum_{j=1}^{m} (\bar{y}_{ij(p)} - \bar{y}_{i(mp)})^2$$

$$\bar{s}_w^2 = \frac{1}{n} \sum_{i=1}^{n} s_i^2$$

$$\bar{\bar{s}}_p^2 = \frac{1}{nm} \sum_{i=1}^{n} \sum_{j=1}^{m} s_{ij}^2$$

and

$$s_{ij}^2 = \frac{1}{p-1} \sum_{k=1}^{p} (y_{ijk} - \bar{y}_{ij(p)})^2.$$

Proof. Clearly

$$E(s_{ij}^2) = S_{ij}^2.$$

Hence, we obtain

$$E(\bar{\bar{s}}_p^2) = E\left[\frac{1}{nm} \sum_{i=1}^{n} \sum_{j=1}^{m} s_{ij}^2\right]$$

$$= \frac{1}{NM} \sum_{i=1}^{N} \sum_{j=1}^{M} S_{ij}^2 = \bar{\bar{S}}_p^2. \tag{8.6.5}$$

Now, s_i^2 is the mean sum of squares between the selected second-stage unit means in the selected i^{th} first-stage unit when the selected i^{th} first-stage unit is fixed, and we have a two-stage sample drawn from it. Hence, for a two-stage sample drawn from the fixed i^{th} first-stage unit, from (8.2.8), we have

$$E(s_i^2) = S_i^2 + \left(\frac{1}{p} - \frac{1}{P}\right) \frac{\sum_{j=1}^{M} S_{ij}^2}{M}. \tag{8.6.6}$$

Hence,

$$\begin{aligned}
E(\bar{s}_w^2) &= E\left(\frac{1}{n}\sum_{i=1}^{n} s_i^2\right) \\
&= E\left[\frac{1}{n}\sum_{i=1}^{n} S_i^2 + \frac{1}{n}\left(\frac{1}{p} - \frac{1}{P}\right)\frac{\sum_{i=1}^{n}\sum_{j=1}^{M} S_{ij}^2}{M}\right] \\
&= \frac{1}{N}\sum_{i=1}^{N} S_i^2 + \left(\frac{1}{p} - \frac{1}{P}\right)\frac{\sum_{i=1}^{N}\sum_{j=1}^{M} S_{ij}^2}{NM} \\
&= \bar{S}_w^2 + \left(\frac{1}{p} - \frac{1}{P}\right)\bar{\bar{S}}_p^2. \qquad (8.6.7)
\end{aligned}$$

Using (8.6.5) in (8.6.7), we obtain

$$E\left(\bar{s}_w^2 - \left(\frac{1}{p} - \frac{1}{P}\right)\bar{\bar{s}}_p^2\right) = \bar{S}_w^2. \tag{8.6.8}$$

Now,

$$(n-1)s^2 = \sum_{i=1}^{n} \bar{y}_{i(mp)}^2 - n\bar{y}_{(nmp)}^2.$$

Hence,

$$
\begin{aligned}
(n-1)E(s_b^2) &= E_1\left[\sum_{i=1}^{n} E_2(\bar{y}_{i(mp)}^2|i)\right] - n\,Var(\bar{y}_{(nmp)}) - n\bar{y}_{...}^2 \\
&= E_1\left[\sum_{i=1}^{n}\{Var(\bar{y}_{i(mp)}|i) + \bar{y}_{i..}^2\}\right] - n\,Var(\bar{y}_{(nmp)}) - n\bar{y}^2\ldots \\
&= E_1\left[\sum_{i=1}^{n}\left\{\left(\frac{1}{m}-\frac{1}{M}\right)S_i^2 + \frac{1}{m}\left(\frac{1}{p}-\frac{1}{P}\right)\frac{\sum_{j=1}^{M} S_{ij}^2}{M} + \bar{y}_{i..}^2\right\}\right] \\
&\quad - n\bar{y}_{...}^2 - n\,Var(\bar{y}_{(nmp)}) \\
&= n\left(\frac{1}{m}-\frac{1}{M}\right)\bar{S}_w^2 + \frac{n}{m}\left(\frac{1}{p}-\frac{1}{P}\right)\bar{\bar{S}}_p^2 \\
&\quad + \frac{n}{N}\sum_{i=1}^{N}\bar{y}_{i..}^2 - n\bar{y}_{...}^2 - n\,Var(\bar{y}_{(nmp)}) \\
&= \frac{n(N-1)}{N}S_b^2 + n\left(\frac{1}{m}-\frac{1}{M}\right)\bar{S}_w^2 + \frac{n}{m}\left(\frac{1}{p}-\frac{1}{P}\right)\bar{\bar{S}}_p^2 \\
&\quad - n\left[\left(\frac{1}{n}-\frac{1}{N}\right)S_b^2 + \frac{1}{n}\left(\frac{1}{m}-\frac{1}{M}\right)\bar{S}_w^2 + \frac{1}{nm}\left(\frac{1}{p}-\frac{1}{P}\right)\bar{\bar{S}}_p^2\right] \\
&= (n-1)S_b^2 + (n-1)\left(\frac{1}{m}-\frac{1}{M}\right)\bar{S}_w^2 + \frac{(n-1)}{m}\left(\frac{1}{p}-\frac{1}{P}\right)\bar{\bar{S}}_p^2.
\end{aligned}
$$

Hence, canceling the factor $(n-1)$, we obtain

$$E(s_b^2) = S_b^2 + \left(\frac{1}{m}-\frac{1}{M}\right)\bar{S}_w^2 + \frac{1}{m}\left(\frac{1}{p}-\frac{1}{P}\right)\bar{\bar{S}}_p^2. \tag{8.6.9}$$

Using (8.6.5), and (8.6.8), from (8.6.9) we obtain

$$E\left[s_b^2 - \left(\frac{1}{m}-\frac{1}{M}\right)\bar{s}_w^2 - \frac{1}{M}\left(\frac{1}{p}-\frac{1}{P}\right)\bar{\bar{s}}_p^2\right] = S_b^2. \tag{8.6.10}$$

Hence, using (8.6.10), (8.6.8) and (8.6.5), we obtain an unbiased estimator of the variance of $\bar{y}_{(nmp)}$ as

$$Est.Var(\bar{y}_{(nmp)}) = \left(\frac{1}{n}-\frac{1}{N}\right)s_b^2 + \frac{1}{N}\left(\frac{1}{m}-\frac{1}{M}\right)\bar{s}_w^2 + \frac{1}{NM}\left(\frac{1}{p}-\frac{1}{P}\right)\bar{\bar{s}}_p^2. \tag{8.6.11}$$

Remark. When N is very large, we have

$$Est.Var(\bar{y}_{(nmp)}) = \frac{1}{n}s_b^2. \tag{8.6.12}$$

8.7 Optimum Values of n, m, and p in Three-stage Sampling

We shall determine the optimum values of n, m and p which will minimize the sampling variance of three-stage sampling for fixed cost of the survey C_0, say. We shall assume that the cost function is of the form

$$C = c_1 n + c_2 nm + c_3 nmp. \tag{8.7.1}$$

Hence, we minimize the variance of $\bar{y}_{(nmp)}$ subject to the restriction

$$C_0 = c_1 n + c_2 nm + c_3 nmp. \tag{8.7.2}$$

Consider the product

$$\begin{aligned} C_0\left[Var(\bar{y}_{(nmp)}) + \frac{S_b^2}{N}\right] &= \left[c_2\left(S_b^2 - \frac{1}{M}\bar{S}_w^2\right)m + \left(\bar{S}_w^2 - \frac{1}{p}\bar{\bar{S}}_p^2\right)\frac{c_1}{m}\right] \\ &\quad + \left[c_3\left(\bar{S}_w^2 - \frac{1}{P}\bar{\bar{S}}_p^2\right)p + \frac{c_2 \cdot \bar{\bar{S}}_p^2}{p}\right] \\ &\quad + \left[c_3\left(S_b^2 - \frac{1}{M}\bar{S}_w^2\right)mp + \frac{c_1\bar{\bar{S}}_p^2}{mp}\right] \\ &\quad + \text{terms independent of } m \text{ and } p. \end{aligned} \tag{8.7.3}$$

When $\Delta_1 = S_b^2 - \frac{1}{M}\bar{S}_w^2$ and $\Delta_2 = \bar{S}_w^2 - \frac{1}{P}\bar{\bar{S}}_p^2$ are both positive, we can write (8.7.3) as

$$\begin{aligned} C_0\left[Var(\bar{y}_{(nmp)}) + \frac{S_b^2}{N}\right] &= \left[\sqrt{c_2 m\Delta_1} - \sqrt{\frac{c_1\Delta_2}{m}}\right]^2 + \left[\sqrt{c_3 p\Delta_2} - \sqrt{\frac{c_2\bar{\bar{S}}_p^2}{p}}\right]^2 \\ &\quad + \left[\sqrt{c_3 mp\Delta_1} + \sqrt{\frac{c_1\bar{\bar{S}}_p^2}{mp}}\right]^2 \\ &\quad + \text{terms independent of } m \text{ and } p. \end{aligned} \tag{8.7.4}$$

Clearly, (8.7.4) is minimum when the three squared terms are all zero. Hence, we obtain the optimum values m and p as

$$\hat{m} = \sqrt{\frac{c_1\Delta_2}{c_2\Delta_1}}, \quad \hat{p} = \sqrt{\frac{c_2\bar{\bar{S}}_p^2}{c_3\Delta_2}}. \tag{8.7.5}$$

The optimum value of n is obtained from $C_0 = c_1 n + c_2 nm + c_3 nmp$ and is given by

$$\hat{n} = \frac{C_0}{c_1 + c_2\hat{m} + c_3\hat{m}\hat{p}}. \tag{8.7.6}$$

The above solutions are obtained only when $\Delta_1 > 0$, and $\Delta_2 > 0$. We now consider other cases for Δ_1 and Δ_2.

Case 1. $\Delta_1 \leq 0$, and $\Delta_2 > 0$.

From (8.7.3), it follows that (8.7.3) is minimum if m assumes the maximum attainable value, say, m, which can be equal to M and p is such that

$$c_3 p\Delta_2 + \frac{c_2\bar{\bar{S}}_p^2}{p} + c_3\hat{m}p\Delta_1 + \frac{1}{\hat{m}p}c_1\bar{\bar{S}}_p^2 \tag{8.7.7}$$

is minimum. If

$$c_3\Delta_2 + c_3\hat{m}\Delta_1 \leq 0, \tag{8.7.8}$$

then (8.7.7) is minimum when p has the maximum attainable value. If (8.7.8) is positive, then the minimum value of p is obtained by equating the partial derivative of (8.7.7) w.r.t. p to zero and is given by

$$\hat{p} = \frac{\sqrt{\left(\frac{c_1}{\hat{m}} + c_2\right)\bar{\bar{S}}_p^2}}{\sqrt{c_3(\Delta_2 + \hat{m}\Delta_1)}}. \tag{8.7.9}$$

Case 2. $\Delta_1 \leq 0$, and $\Delta_2 \leq 0$.

In this case (8.7.3) is minimum when p has the maximum attainable integral value, which could also be equal to P, and m is such that

$$c_2\Delta_1 m + \Delta_2\frac{c_1}{m} + m\hat{p}c_3\Delta_1 + \frac{c_1\bar{\bar{S}}_p^2}{m\hat{p}} \tag{8.7.10}$$

is minimum. Now, (8.7.10) can be written as

$$m(c_2 + \hat{p}c_3)\Delta_1 + \frac{c_1}{m}(\Delta_2) + \frac{c_1}{m\hat{p}}\bar{\bar{S}}_p^2$$

which is minimum when m has the maximuum attainable integral value.

Case 3. $\Delta_1 > 0$ and $\Delta_2 \leq 0$.

In this case, the solution for pm is given by

$$\hat{p}\hat{m} = \sqrt{\frac{c_1\bar{\bar{S}}_p^2}{c_2\Delta_1}}$$

with p and m attaining maximum possible integer.

EXERCISES

8.1. To estimate the total number of words in an English dictionary, 10 out of 26 alphabets were selected with SRSWOR. From the pages assigned to selected alphabets, pages were selected with SRSWOR and the number of words were counted. The data are given in Table 8.

Table 8.1

Sr. No	Sample alphabet	No. of pages M_i	No. of words in sample page 1	2
1	B	49	34	40
2	S	131	54	60
3	C	97	39	50
4	N	21	44	20
5	P	85	20	35
6	F	43	30	18
7	U	18	12	16
8	T	7	25	14
9	X	49	50	40
10	E	54	60	45

(Total No. of pages in the dictionary is 1000)

(a) Estimate unbiasedly the total number of words in the dictionary and obtain estimate of its variance.

(b) Estimate the efficiency of the method of sampling compared to that of drawing 20 pages from the dictionary with SRSWOR.

8.2. A population consists of N clusters of M units each. It is proposed to draw a sample of n clusters with SRSWOR from the N clusters and from each selected cluster, a subsample of m units is to be selected with SRSWOR for estimating the mean per unit in the population. Suppose that the cost function is given by $C = c_0 + c_1 n + c_2 m$. For fixed cost $C_0 = \$1000$, and given that $c_0 = \$300$, $c_1 = \$9$, $c_2 = \$1$, determine the optimum values of n and m, using the following analysis of variance table.
($N = 90$, $M = 20$).

8.3. A sample of 14 blocks is selected with *pp* to census population from a region. Households M_i within selected blocks are listed, from which a SRSWOR of m_i households is selected and information on monthly expenditure on vegetables y is collected. The data are given in

Table 8.2 Analysis of variance table.

Source	Sum of Squares	df	Mean SS
Between Clusters	$20\sum_{i=1}^{90}(\bar{y}_{i.} - \bar{y}_{..})^2$	89	180.9
Within Clusters	$\sum_{i=1}^{90}\sum_{j=1}^{20}(y_{ij} - \bar{y}_{i.})^2$	1710	49.5
Total	$\sum_{i=1}^{90}\sum_{j=1}^{20}(y_{ij} - \bar{y}_{..})^2$	1799	56.0

Table 8.3

P_i	M_i	m_i	y_i	P_i	M_i	m_i	y_i
0.0440	78	10	110	0.0197	50	6	80
0.0291	63	8	205	0.0334	80	10	130
0.0596	160	20	250	0.0930	120	16	200
0.0930	124	16	200	0.0499	105	15	250
0.0052	40	6	29	0.0370	110	15	110
0.0741	142	18	140	0.0510	90	12	200
0.0374	110	15	110	0.0725	145	20	275

Table 8.4

Stratum	Area under wheat	Village 1		Village 2	
		Plot 1	Plot 2	Plot 1	Plot 2
1	2800	32	245	125	43
2	3805	30	140	170	113
4	6640	220	275	60	183
5	7215	45	233	210	190
6	2900	160	117	177	165
7	2775	95	110	350	259
8	1590	80	67	85	110
9	3504	370	389	325	311
10	1459	90	185	143	171

Table 8.3. Obtain the per household monthly expenditure on vegetables and compute the standard error of the estimate.

8.4. The villages in a population are allocated to 10 strata and a field-to-field complete enumeration is made to find the total area under wheat. From each stratum two villages are selected with replacement and with *pp* to the number of fields growing wheat. From a selected village, a SRSWOR of 2 fields growing wheat is taken and a plot of a given size and shape is selected at random in the field. The plot yields are given in Table 8.4. Using the unweighted mean in the stratum, find the yield per plot and its standard error.

Table 8.5

Block	1	2	3	4	5	6	7	8	9	10
Rooms	40	60	55	58	70	53	60	58	50	75
Persons	90	110	99	80	110	110	100	120	100	140

8.5. A city block is divided into 100 blocks from which 10 blocks are selected with replacement and with *pp* to the number of households enumerated in a previous census. A sampling fraction of 5 percent is used for selecting a SRSWOR of households from each block. The sample gives the data on the number of rooms and the number of persons in them. The data are given in Table 8.5. Estimate the number of persons per room and compute the standard error of the estimate.

8.6. It is proposed to estimate the average household expenditure per week on durable goods in a region containing 400 villages. A sample of n villages is selected with replacement and with *pp* to the number of households in the village. From a selected village m households are selected with SRSWOR and information on monthly expenditure on durable goods is obtained. The variance of the sample mean is given by

$$Var(\bar{y}) = \frac{128}{n} + \frac{1500}{nm}.$$

The cost function is known to be $c = 2n + nm$. For given cost of \$350, determine the values of n and m which will minimize the variance.

8.7. A population contains N first-stage units, the ith first stage unit having M_i second-stage units, and let $M_0 = \sum_{i=1}^{N} M_i$. A sample of n fsu's is selected with replacement and with probabilities $\pi_i = \frac{M_i}{M_0}$. If a fsu occurs in the sample m times, a WOR random sample of $m\lambda$ ssu's is selected from it. Prove that $t_1 = \sum_{i=1}^{N} \frac{\lambda_i \bar{y}_i}{n}$ is an unbiased estimator of $\theta = \sum_{i=1}^{N} \pi_i \theta_i$ where θ_i is the mean per ssu in the ith fsu and λ_i is the frequency with which this fsu occurs in the sample, $\bar{y}_i$ being defined as 0 when no sample is taken from the fsu. Further, prove that

$$Var(t_1) = \frac{1}{n}\sum_{i=1}^{N} \pi_i \left(\frac{1}{m} - \frac{1}{M_i}\right) S_i^2 + \frac{1}{n}\sum_{i=1}^{N} \pi_i(\theta_i - \theta)^2 - \frac{1}{n}(n-1)\sum_{i=1}^{N} \frac{\pi_i^2 S_i^2}{M_i},$$

where $S_i^2 = \sum_{j=1}^{M_i} \frac{(y_{ij} - \theta_i)^2}{(M_i - 1)}$.

If subsampling is carried out independently every time a fsu enters the sample, taking a WOR random sample of m ssu's, prove that $t_2 = \sum_{i=1}^{n} \frac{\bar{y}_i}{n}$ is an unbiased estimator of θ and that

$$Var(t_2) = \frac{1}{n}\sum_{i=1}^{N} \pi_i \left(\frac{1}{m} - \frac{1}{M_i}\right) S_i^2 + \frac{1}{n}\sum_{i=1}^{N} \pi_i(\theta_i - \theta)^2.$$

Hence, show that $Var(t_1) < Var(t_2)$. Show that unbiased estimators of variances of t_1 and t_2 are given by

$$Est.Var(t_1) = \frac{\sum_{i=1}^{n} \lambda_i(\bar{y}_i - \bar{y})^2}{n(n-1)} - \frac{\sum_{i=1}^{n} s_i^2}{n(n-1)}\left[(n-1)\lambda_i \frac{\pi_i}{M_i} - \frac{\lambda_i - 1}{m}\right]$$

$$Est.Var(t_2) = \frac{\sum_{i=1}^{n} (\bar{y}_i - \bar{y})^2}{n(n-1)}.$$

8.8. In two-stage sampling, one ssu is selected with *pp* to x from the whole population. If this happens to come from the ith fsu, a WOR random sample of $m_i - 1$ ssu's is taken from the remaining $M_i - 1$ ssu's left in i^{th} fsu. From the other $(N-1)$ fsu's a WOR random sample of $(n-1)$ fsu's is selected. Subsampling of the selected fsu's is SRSWOR. Show that $\frac{\sum_{i=1}^{n} M_i \bar{y}_i}{\sum_{i=1}^{n} M_i \bar{x}_i}$ is an unbiased estimator of $R = \frac{Y}{X}$.

Chapter 9

DOUBLE SAMPLING

9.1 Double Sampling

We have seen in Chapters 5 and 6 the use of auxiliary information on the variable x for improving the efficiency of the estimator of the population mean of the study variable y. But suppose that information on x is not available. Then, in this case, there are two courses open to us. We may decide to select a sample of n units, collect information on y, and use the sample mean $\bar{y}$ as an estimator of the population mean $\bar{Y}$. Alternatively, we may use a part of the budget for collecting information on x for a fairly large sampe of size n' units taken from the population and then select a smaller sample of size n units from it and collect information on y. These two samples are used to obtain a better estimator of the population mean for y. The second procedure is called double sampling or two-phase sampling. It is useful when it is considerably cheaper and quicker to collect data on x than on y and when there is high correlation between x and y. The procedure of selecting a large sample for collecting information of auxiliary variable x and then selecting a sub-sample from it for collecting the information on the study variable y is called double sampling.

Let us illustrate the usefulness of double sampling by a numerical example. Consider the data of Table 9.1 about the number of workers employed in 10 factories, where x and y denote the numbers of workers employed in 1975 and in 1979 respectively.

Suppose for costs \$60 are available to us. Then, we select a SRSWOR of $n = 3$ factories and estimate $\bar{y}$ from it. The sampling variance for size $n = 3$ is easily verified to be

$$Var(\bar{y}) = 74.56. \tag{9.1.1}$$

Table 9.1 Number of workers employed by ten factories.

Factory	No. of Workers in 1975 x	No. of Workers in 1979 y
1	30	31
2	15	15
3	60	67
4	18	20
5	25	27
6	45	48
7	10	9
8	20	22
9	12	13
10	15	18
Total	250	270

Suppose it costs \$2 per factory to collect information on x which is obtained by mail. Then, we spend \$20 to collect information on x for all the ten factories, and then we select a sample of 2 units from it and collect information on y, and thus spend another \$40, thus making a total of \$60 available to us. We now use difference estimator $\bar{y}_d = \bar{y} - \bar{x} + \bar{X}$ as an estimator of the population mean $\bar{Y}$. One can easily verify that

$$Var(\bar{y}_d) = 1.87 \tag{9.1.2}$$

and hence the efficiency of the double sampling is obtained as

$$Eff. = \frac{74.56}{1.87} \times 100 = 398.7\%.$$

We thus see that it is advantageous to use double sampling.

Double sampling may be used for introducing a desired stratification or making a good estimate of X and $\bar{X}$ for the purpose of ratio, regression or difference estimation.

9.2 Double Sampling for Ratio Estimation

The usual ratio estimator $\bar{y}_r = \frac{\bar{y}}{\bar{x}}\bar{X}$ cannot be used if $\bar{X}$ is unknown. In this case, double sampling is used. An initial sample of size n' is selected with SRSWOR and information on x is collected. Let $\bar{x}_{n'}$ be the mean of x in the initial sample. Then, a second sample is a subsample of size n selected from the initial sample with SRSWOR. Let $\bar{y}$ and $\bar{x}$ denote the means of y and x for the second sample. Then, the following ratio estimator is suggested

$$\bar{y}_{nd} = \frac{\bar{y}_n}{\bar{x}_n}\bar{x}_{n'}. \tag{9.2.1}$$

In the following theorems, we derive the bias and approximate (mean square error) for the estimator (9.2.1).

Theorem 9.2.1. *The relative bias of $\bar{y}_{nd}$ is given by*

$$\left(\frac{1}{n}-\frac{1}{n'}\right)(c_x^2 - \rho c_x c_y)$$

where c_x, c_y, and ρ denote respectively the coefficients of variations of x, y and correlation coefficient between x and y.

Proof. Let

$$\psi_0 = \frac{\bar{y}_n - \bar{Y}_N}{\bar{Y}_N}, \quad \psi_1 = \frac{\bar{x}_n - \bar{X}_N}{\bar{X}_N}, \quad \psi_2 = \frac{\bar{x}_{n'} - \bar{X}_N}{\bar{X}_N}. \tag{9.2.2}$$

Then, clearly, we have

$$E(\psi_0) = E(\psi_1) = E(\psi_2) = 0 \tag{9.2.3}$$

$$\begin{aligned} E(\psi_0^2) &= \frac{Var(\bar{y}_n)}{\bar{Y}_N^2} \\ &= \frac{1}{\bar{Y}_N^2}[V_1 E_2(\bar{y}_n|n') + E_1 V_2(\bar{y}_n|n')] \\ &= \frac{1}{\bar{Y}_N^2}\left[V_1(\bar{y}_{n'}) + E_1\left\{\left(\frac{1}{n}-\frac{1}{n'}\right)s_y'^2\right\}\right] \\ &= \frac{1}{\bar{Y}_N^2}\left[\left(\frac{1}{n'}-\frac{1}{N}\right)S_y^2 + \left(\frac{1}{n}-\frac{1}{n'}\right)S_y^2\right] \\ &= \frac{1}{\bar{Y}_N^2}\left(\frac{1}{n}-\frac{1}{N}\right)S_y^2 \end{aligned} \tag{9.2.4}$$

where $s_y'^2$ = mean sum of squares of y in the initial sample.

Theorem 9.2.1. *The relative bias of $\bar{y}_{nd}$ is given by*

$$E(\psi_1^2) = \frac{Var(\bar{x}_n)}{\bar{X}_N^2} = \frac{1}{\bar{X}_N^2}\left(\frac{1}{n}-\frac{1}{N}\right)S_x^2. \tag{9.2.5}$$

It follows on the same lines as in (9.2.4). Clearly

$$E(\psi_2^2) = \frac{\left(\frac{1}{n'}-\frac{1}{N}\right)S_x^2}{\bar{X}_N^2}. \tag{9.2.6}$$

Next,

$$\begin{aligned}
E(\psi_0\psi_1) &= \frac{1}{\bar{X}_N\bar{Y}_N} Cov(\bar{x}_n, \bar{y}_n) \\
&= \frac{1}{\bar{X}_N\bar{Y}_N}[Cov\{E(\bar{x}_n|n'), E(\bar{y}_n|n')\} + E\{Cov(\bar{x}_n, \bar{y}_n)|n'\}] \\
&= \frac{1}{\bar{X}_N\bar{Y}_N}\left[Cov(\bar{x}_{n'}, \bar{y}_{n'}) + E\left\{\left(\frac{1}{n} - \frac{1}{n'}\right)s'_{yx}\right\}\right] \\
&= \frac{1}{\bar{X}_N\bar{Y}_N}\left[\left(\frac{1}{n'} - \frac{1}{N}\right)S_{yx} + \left(\frac{1}{n'} - \frac{1}{N}\right)S_{yx}\right] \\
&= \frac{1}{\bar{X}_N\bar{Y}_N}\left(\frac{1}{n'} - \frac{1}{N}\right)S_{yx} \qquad (9.2.7)
\end{aligned}$$

when s'_{yx} is the mean sum of products in the initial sample. Further,

$$\begin{aligned}
E(\psi_0\psi_2) &= \frac{Cov(\bar{y}_n, \bar{x}_{n'})}{\bar{Y}_N\bar{X}_N} \\
&= \frac{1}{\bar{Y}_N\bar{X}_N}[Cov\{E(\bar{y}_n|n'), E(\bar{x}_{n'}|n')\} + E\{Cov(\bar{y}_n, \bar{y}_{n'})|n'\}] \\
&= \frac{1}{\bar{Y}_N\bar{X}_N}[Cov(\bar{y}_{n'}, \bar{x}_{n'}) + E\{0\}] \\
&= \frac{1}{\bar{Y}_N\bar{X}_N}\left(\frac{1}{n'} - \frac{1}{N}\right)S_{yx} \qquad (9.2.8)
\end{aligned}$$

and

$$\begin{aligned}
E(\psi_1\psi_2) &= \frac{Cov(\bar{x}_n, \bar{x}_{n'})}{\bar{X}_N^2} \\
&= \frac{1}{\bar{X}_N^2}[Cov\{E(\bar{x}_n|n'), E(\bar{x}_{n'}|n')\} + 0] \\
&= \frac{1}{\bar{X}_N^2} Var(\bar{X}_{n'}). \qquad (9.2.9)
\end{aligned}$$

We can now write $\bar{y}_{nd}$ as

$$\begin{aligned}
\bar{y}_{nd} &= \frac{\bar{Y}_N(1+\psi_0)}{\bar{X}_N(1+\psi_1)}\bar{X}_N(1+\psi_2) \\
&= \bar{Y}_N(1+\psi_0)(1+\psi_2)(1+\psi_1)^{-1} \\
&= \bar{Y}_N(1+\psi_0+\psi_2+\psi_0\psi_2)(1-\psi_1+\psi_1^2\ldots) \\
&= \bar{Y}_N(1-\psi_1+\psi_1^2+\psi_0-\psi_0\psi_1+\psi_2-\psi_2\psi_1+\psi_0\psi_2), \qquad (9.2.10)
\end{aligned}$$

since terms involving (ψ_0, ψ_1, ψ_2) of degree greater than two are assumed to be negligible. Taking expectation of (9.2.10), we obtain

$$E(\bar{y}_{nd}) = \bar{Y}_N(1 + E(\psi_1^2) - E(\psi_0\psi_)) - E(\psi_2\psi_1) + E(\psi_0\psi_2)). \quad (9.2.11)$$

Substituting from (9.2.5), (9.2.7), (9.2.9), (9.2.8) in (9.2.11), we obtain

$$E(\bar{y}_{nd}) = \bar{Y}_N \left[1 + \left(\frac{1}{n} - \frac{1}{n'}\right)(C_x^2 - \rho C_x C_y)\right]. \quad (9.2.12)$$

Hence, relative bias of $\bar{y}_{nd}$ is given by

$$\frac{E(\bar{y}_{nd}) - \bar{Y}_N}{\bar{Y}_N} = \left(\frac{1}{n} - \frac{1}{n'}\right)(C_x^2 - \rho C_x C_y). \quad (9.2.13)$$

Theorem 9.2.2. *The mean square error or approximate variance of* $\bar{y}_{nd}$ *is given by*

$$MSE(\bar{y}_{nd}) = \left(\frac{1}{n} - \frac{1}{n'}\right)(S_y^2 + R^2S_x^2 - 2RS_{yx}) + \left(\frac{1}{n'} - \frac{1}{N}\right)S_y^2,$$

where $R = \frac{\bar{Y}_N}{\bar{X}_N}$.

Proof. The $MSE(\bar{y}_{nd})$ is given by

$$\begin{aligned} MSE(\bar{y}_{nd}) &= E(\bar{y}_{nd} - \bar{Y}_N)^2 \\ &= \bar{Y}_N^2 E[-\psi_1 + \psi_0 + \psi_2]^2 \\ &= Y_N[\psi_0^2 + \psi_1^2 + \psi_2^2 - 2\psi_0\psi_1 + 2\psi_0\psi_2 - 2\psi_1\psi_2], \end{aligned} \quad (9.2.14)$$

since terms in (ψ_0, ψ_1, ψ_2) of degree greater than two are negligible. Substituting from (9.2.4), (9.2.5), (9.2.6), (9.2.7), (9.2.8) and (9.2.9) in (9.2.14), we obtain (9.2.15)

$$MSE(\bar{y}_{nd}) = \left(\frac{1}{n} - \frac{1}{n'}\right)(S_y^2 + R^2S_x^2 - 2RS_{yx}) + \left(\frac{1}{n'} - \frac{1}{N}\right)S_y^2. \quad (9.2.15)$$

Corollary 9.1.1. *The estimator* $\bar{y}_{nd}$ *based on double sampling is more efficient than the estimator* $\bar{y}_n$ *based on SRSWOR when no auxiliary variable is used if*

$$\rho \geq \frac{1}{2}\frac{C_x}{C_y}.$$

9.3 Double Sampling for Difference Estimation

An initial sample of size n' is selected witih SRSWOR from a population of size N and information on x is obtained. Then, a second sample of size n is selected from the initial sample with SRSWOR and information on y is measured on it. Let $\bar{y}_n$, $\bar{x}_n$ denote the sample means for the second sample, while $\bar{x}_{n'}$ denotes the sample mean of the initial sample. Let k be a given estimate of the ratio $R = \frac{\bar{Y}_N}{\bar{X}_N}$ in the population. Then, the following difference estimator is used for estimating $\bar{Y}_N$.

$$\bar{y}_d = \bar{y}_n - k\bar{x}_n + k\bar{x}_{n'} \tag{9.3.1}$$

Now,

$$\begin{aligned} E(\bar{y}_d) &= E(\bar{y}_n - k\bar{x}_n + k\bar{x}_{n'}) \\ &= E_1[E_2\{(\bar{y}_n - k\bar{x}_n + k\bar{x}_{n'})|n'\}] \\ &= E_1[\bar{y}_{n'} - k\bar{x}_{n'} + k\bar{x}_{n'}] \\ &= E_1(\bar{y}_{n'}) = \bar{Y}_N, \end{aligned} \tag{9.3.2}$$

which shows that the estimator $\bar{y}_d$ is unbiased for $\bar{Y}_N$. Further,

$$\begin{aligned} Var(\bar{y}_d) &= Var(\bar{y}_n - k\bar{x}_n + k\bar{x}_{n'}) \\ &= V_1E_2[(\bar{y}_n - k\bar{x}_n + k\bar{x}_{n'})|n'] + E_1V_2[(\bar{y}_n - k\bar{x}_n + k\bar{x}_{n'})|n'] \\ &= V_1E_2(\bar{y}_{n'}) + E_1\left[\left(\frac{1}{n} - \frac{1}{n'}\right)(s_y'^2 + k^2s_x'^2 - 2ks_{yx})\right], \end{aligned} \tag{9.3.3}$$

where $s_y'^2$, $s_x'^2$, and s_{yx} refer to the quantities for the initial sample. Next (9.3.3) can be written as

$$Var(\bar{y}_d) = \left(\frac{1}{n} - \frac{1}{N}\right)S_y^2 - \left(\frac{1}{n} - \frac{1}{n'}\right)kS_x(2\rho S_y - kS_x), \tag{9.3.4}$$

where ρ is the correlation coefficient between y and x.

Let c' and c denote the unit costs of collecting the information on x and y respectively. Clearly, c' will be much smaller than c. Then, the total cost of double sampling procedure is

$$C = c'n' + cn. \tag{9.3.5}$$

If a SRSWOR is taken (without using double sampling), then the sample size for the same cost would be

$$n_0 = \frac{c'n' + cn}{c} = n + \frac{c'n'}{c}$$

and the variance of the mean of this sample would be

$$Var(\bar{y}_{n_0}) = \left(\frac{1}{n_0} - \frac{1}{N}\right) S_y^2. \tag{9.3.6}$$

Then double sampling will be more efficient than the SRSWOR if no information on x is used for the same cost, if

$$2\rho - h > \frac{1}{\left[\left(1 - \frac{n}{n'}\right)\left(1 + \frac{nc}{n'c'}\right)\right]} \tag{9.3.7}$$

where $h = k\frac{S_x}{S_y}$.

As an example, let k be the regression coefficient $\rho\frac{S_y}{S_x}$, then $h = \rho$ and let $\frac{n}{n'} = .1$, $\frac{nc}{n'c'} = 9$. Then, the above condition (9.3.7) becomes $\rho > \frac{1}{9}$.

9.4 Double Sampling for Regression Estimation

The first sample is a SRSWOR of size n' from the population of size N and the second sample is a subsample of size n drawn from the first sample with SRSWOR. Let $\bar{y}_n$, $\bar{x}_n$, b_n be the sample means of x and y, and the sample regression coefficient of y on x for the second sample. Let $\bar{x}_{n'}$ be the sample mean of x for the first sample. Then the double-sampling regression estimator of $\bar{Y}$ is given by

$$\bar{y}_{dr} = \bar{y}_n - b_n(\bar{x}_n - \bar{x}_{n'}). \tag{9.4.1}$$

We shall obtain a large sample approximate variance of $\bar{y}_{dr}$. We shall write $\bar{x}_{n'}$ as $\lambda\bar{x}_n + \mu\bar{x}'$, where $\mu = 1 - \lambda$ and $\bar{x}'$ is the mean of the $n'\mu$ units in the first sample not common with the second sample. Then, we can write y_{dr} as $\bar{y}_{dr} = \bar{y}_n - b_n(\bar{x}_n - \bar{x}')$. If n is large, $\bar{y}_{dr}$ will be distributed as

$$\bar{y}_n - \mu\beta(\bar{x}_n - \bar{x}'), \tag{9.4.2}$$

when $\beta = \rho\frac{S_y}{S_x}$ = the population regression coefficient of y on x. Hence, the large-sample variance of $\bar{y}_{dr}$ is given by

$$\begin{aligned}
Var(\bar{y}_{dr}) &= \frac{S_y^2}{n} + \mu^2 B^2 \left(\frac{1}{n} + \frac{1}{n'}\right) S_x^2 - 2\mu B \frac{\rho S_y S_x}{n} \\
&= \frac{S_y^2}{n}\left[1 + \mu^2\rho^2\left(1 + \frac{n}{n'\mu}\right) - 2\mu\rho^2\right] \\
&= \frac{S_y^2}{n}\left[1 + \mu^2\rho^2 + \frac{n}{n'}\mu\rho^2 - 2\mu\rho^2\right] \\
&= \frac{S_y^2}{n}[1 + \mu^2\rho^2 - \mu\rho^2 + \lambda\mu\rho^2 - \mu\rho^2] \\
&= \frac{S_y^2}{n}[1 - \mu\lambda\rho^2 + \lambda\mu\rho^2 - \mu\rho^2] \\
&= \frac{S_y^2}{n}[1 - (1-\lambda)\rho^2] \\
&= \frac{S_y^2}{n} - \frac{1}{n}\left(1 - \frac{n}{n'}\right)\rho^2 S_y^2 \\
&= \frac{S_y^2(1-\rho^2)}{n} + \frac{\rho^2 S_y^2}{n'}
\end{aligned} \tag{9.4.3}$$

which shows that double sampling is more precise than a direct SRSWOR of size n.

Let the cost function be given by

$$c = c'n' + cn$$

and the cost of the survey be fixed at C_0. Then,

$$C_0 = c'n' + cn. \tag{9.4.4}$$

We shall now determine the optimum value of the subsampling rate $\frac{n}{n'}$ which will minimize the variance (9.4.3) subject to (9.4.4). We shall therefore minimize

$$\phi = \frac{S_y^2}{n}\left[1 - \left(1 - \frac{n}{n'}\right)\rho^2\right] - L(c'n' + cn - C_0). \tag{9.4.5}$$

where L is a Lagrange multiplier. Equating the partial derivatives of (9.4.5) w.r.t. n and n' to zero, we get

$$\frac{S_y^2(1-\rho^2)}{n^2} = Lc \tag{9.4.6}$$

$$\frac{S_y^2\rho^2}{n'^2} = Lc'. \tag{9.4.7}$$

Dividing (9.4.7) by (9.4.6), we obtain the optimum value of $\frac{n}{n'}$ as

$$\left(\frac{n}{n'}\right)_{opt} = \sqrt{\frac{c'(1-\rho^2)}{c\rho^2}}. \tag{9.4.8}$$

Then, the minimum variance of the double sampling is given by

$$Var(\bar{y}_{dr})_{min} = \frac{S_y^2}{C_0}\,[\rho\sqrt{c'} + \sqrt{(1-\rho^2)c}]^2. \tag{9.4.9}$$

For the same cost C_0, if a SRSWOR is taken without auxiliary variable, then its size would be

$$n_0 = \frac{C_0}{c}$$

and its sampling variance is

$$Var(\bar{y})_{SRS} = \frac{cS_y^2}{C_0}. \tag{9.4.10}$$

Comparing (9.4.9) and (9.4.10), we see that double sampling with optimum sub-sampling rate is more efficient than the corresponding SRSWOR without auxiliary variable if

$$\rho^2 > \frac{4cc'}{c+c'}. \tag{9.4.11}$$

9.5 Double Sampling for pps Estimation

Suppose it is desired to select the sample with *pp* to x but information on x is not available. Then, in this situation we may use double sampling. An initial sample of size n' is selected with SRSWOR from a population of size N, and information on x is collected for this sample. Then a second sample of size n is selected with replacement and with *pp* to x from the initial sample of size n'. Let $\bar{x}_{n'}$ denote the mean of x for the initial sample. Let $\bar{x}_n$, $\bar{y}_n$ denote means respectively of x and y for the second sample. Then, we have the following theorem.

Theorem 9.5.1.

(i) *An unbiased estimator of the population mean $\bar{Y}$ is given*

$$\hat{\bar{Y}} = \frac{x'}{n'n}\sum_{i=1}^{n}\left(\frac{y_i}{x_i}\right),$$

where x' denotes the total for x in the first sample.

(ii) $Var(\hat{\hat{Y}}) = \left(\frac{1}{n'} - \frac{1}{N}\right) S_y^2 + \frac{(n'-1)}{N(N-1)nn'} \sum_{i=1}^{N} \frac{x_i}{X} \left(\frac{y_i}{\frac{x_i}{X}} - Y\right)^2$, *where* X *and* Y *denote the totals of* x *and* y *respectively in the population.*

(iii) *An unbiased estimator of the variance of* $\hat{\hat{Y}}$ *is given by*

$$Est.Var(\hat{\hat{Y}}) = \left(\frac{1}{n'} - \frac{1}{N}\right) \frac{1}{n(n'-1)} \left[x' \sum_{i=1}^{n} \frac{y_i^2}{x_i} - \frac{x'^2(A-B)}{n'(n-1)}\right]$$
$$+ \frac{1}{n(n-1)} \sum_{i=1}^{n} \left(\frac{x' y_i}{n' x_i} - \hat{\hat{Y}}\right)^2$$

where $A = \left(\sum_{i=1}^{n} \frac{y_i}{x_i}\right)^2$ and $B = \sum_{i=1}^{n} \frac{y_i^2}{x_i^2}$.

Proof. Let E_2 denote the expectation of $\hat{\hat{Y}}$, when the first sample is fixed. The second sample is selected with *pp* to x, hence using Theorem 2.2.1 with $P_i = \frac{x_i}{x'}$, we find that

$$\begin{aligned} E_2\left(\frac{\hat{\hat{Y}}}{n'}\right) &= E_2\left[\frac{1}{n} \sum_{i=1}^{n} \frac{y_i}{n' \frac{x_i}{x'}}\right] \\ &= E_2\left[\frac{x'}{nn'} \sum_{i=1}^{n} \left(\frac{y_i}{x_i}\right)\right] \\ &= \bar{y}_{n'}, \end{aligned} \tag{9.5.1}$$

where $\bar{y}_{n'}$ is the mean of y for the first sample. Hence,

$$\begin{aligned} E(\hat{\hat{Y}}) &= E_1\left[E_2\left(\frac{\hat{\hat{Y}}}{n'}\right)\right] \\ &= E_1(\bar{y}_{n'}) = \hat{\bar{Y}}, \end{aligned}$$

which proves part (i) of the theorem. Further,

$$\begin{aligned} Var(\hat{\hat{Y}}) &= V_1 E_2\left(\frac{\hat{\hat{Y}}}{n'}\right) + E_1 V_2\left(\frac{\hat{\hat{Y}}}{n'}\right) \\ &= V_1(\bar{y}_{n'}) + E_1 V_2\left(\frac{\hat{\hat{Y}}}{n'}\right) \\ &= \left(\frac{1}{n'} - \frac{1}{N}\right) S_y^2 + E_1 V_2\left(\frac{\hat{\hat{Y}}}{n'}\right). \end{aligned} \tag{9.5.2}$$

Now, using Theorem 2.2.1, we get

$$V_2\left(\frac{\hat{\hat{Y}}}{n'}\right) = \frac{1}{nn'^2}\sum_{i=1}^{n'}\frac{x_i}{x'}\left(\frac{y_i}{\frac{x_i}{x'}} - y'\right)^2$$

$$= \frac{1}{nn'^2}\sum_{i=1}^{n'}\sum_{i<j}^{n'} x_i x_j\left(\frac{y_i}{x_i} - \frac{y_j}{x_j}\right)^2, \tag{9.5.3}$$

and hence

$$E_1 V_2\left(\frac{\hat{\hat{Y}}}{n'}\right) = \frac{1}{nn'^2}\frac{n'(n'-1)}{N(N-1)}\sum_{i=1}^{N}\sum_{i<j}^{n'} x_i x_j\left(\frac{y_i}{x_i} - \frac{y_j}{x_j}\right)^2, \tag{9.5.4}$$

since the probability of a specified pair of units being selected in the sample is $\frac{n'(n'-1)}{N(N-1)}$. (9.5.4) can also be expressed as

$$E_1 V_2\left(\frac{\hat{\hat{Y}}}{n'}\right) = \frac{1}{nn'}\frac{(n'-1)}{N(N-1)}\sum_{i=1}^{N}\frac{x_i}{X}\left(\frac{y_i}{\frac{x}{X}} - Y\right)^2. \tag{9.5.5}$$

Substitution from (9.5.5) in (9.5.3) gives

$$Var(\hat{\hat{Y}}) = \left(\frac{1}{n'} - \frac{1}{N}\right)S_y^2 + \frac{(n'-1)}{nn'N(N-1)}\sum_{i=1}^{N}\frac{x_i}{X}\left(\frac{y_i}{\frac{x_i}{X}} - Y\right)^2. \tag{9.5.6}$$

This proves the second part of the theorem.

We now consider the estimation of $Var(\hat{\hat{Y}})$. Given the first sample, we obtain

$$E_2\left[\frac{1}{n}\sum_{i=1}^{n}\frac{y_i^2}{P_i}\right] = \sum_{i=1}^{n'} y_i^2, \tag{9.5.7}$$

where $P_i = \frac{x_i}{x'}$. Also, given the first sample,

$$E_2\left[\frac{1}{n(n-1)}\sum_{i=1}^{n}\left(\frac{y_i}{n'P_i} - \hat{Y}\right)^2\right] = V_2(\hat{Y}) = E_2(\hat{Y}^2) - \bar{y}_{n'}^2.$$

Hence

$$E_2\left[\hat{Y}^2 - \frac{1}{n(n-1)}\sum_{i=1}^{n}\left(\frac{y_i}{n'P_i} - \hat{Y}\right)^2\right] = \bar{y}_{n'}^2. \tag{9.5.8}$$

Substituting $\hat{Y} = \frac{x'}{n'n}\sum_{i=1}^{n}\left(\frac{y_i}{x_i}\right)$, and $P_i = \frac{x_i}{x'}$, (9.5.8) can be rewritten as

$$E_2\left[\frac{x'^2}{nn'^2(n-1)}\left\{\left(\sum_{i=1}^{n}\frac{y_i}{x_i}\right)^2 - \left(\sum_{i=1}^{n}\frac{y_i^2}{x_i^2}\right)\right\}\right] = \bar{y}_{n'}^2. \tag{9.5.9}$$

Using (9.5.7) and (9.5.9), we get

$$E_2\left[\frac{1}{n}\sum_{i=1}^{n} y_i^2\frac{x'}{x_i} - \frac{x'^2}{nn'(n-1)}(A-B)\right] = \sum_{i=1}^{n'} y_i^2 - n'\bar{y}_{n'}^2, \qquad (9.5.10)$$

where $A = \left(\sum_{i=1}^{n}\frac{y_i}{x_i}\right)^2$, and $B = \sum_{i=1}^{n}\frac{y_i^2}{x_i^2}$. From (9.5.10), one obtains

$$E_2\left[\frac{1}{n(n'-1)}\left\{x'\sum_{i=1}^{n}\frac{y_i^2}{x_i} - \frac{x'^2(A-B)}{n'(n-1)}\right\}\right] = s_y'^2,$$

where $s_y'^2$ is the mean sum of squares of y for the first sample. Thus, we obtain

$$E_1E_2\left[\frac{1}{n(n'-1)}\left\{x'\sum_{i=1}^{n}\frac{y_i^2}{x_i} - \frac{x'^2(A-B)}{n'(n-1)}\right\}\right] = E_1(s_y'^2) = S_y^2. \qquad (9.5.11)$$

Note that (9.5.11) gives an unbiased estimator of S_y^2. Next from (9.5.5), we see that

$$E_1V_2\left(\frac{\hat{\hat{Y}}}{n'}\right) = \frac{(n'-1)}{nn'N(N-1)}\sum_{i=1}^{N}\frac{x_i}{X}\left(\frac{y_i}{\frac{x_i}{X}} - Y\right)^2,$$

and from Theorem 2.2.1, we obtain

$$E_2\left[\frac{1}{n(n-1)}\sum_{i=1}^{n}\left(\frac{y_ix'}{n'x_i} - \hat{\hat{Y}}\right)^2\right] = V_2\left(\frac{\hat{\hat{Y}}}{n'}\right).$$

Thus

$$E_1E_2\left[\frac{1}{n(n-1)}\sum_{i=1}^{n}\left(\frac{x'y_i}{n'x_i} - \hat{\hat{Y}}\right)^2\right] = \frac{(n'-1)}{nn'N(N-1)}\sum_{i=1}^{N}\frac{x_i}{X}\left(\frac{y_i}{\frac{x_i}{X}} - Y\right)^2. \qquad (9.5.12)$$

Then (9.5.12) gives an unbiased estimator of

$$\frac{(n'-1)}{nn'N(N-1)}\sum_{i=1}^{N}\frac{x_i}{X}\left(\frac{y_i}{\frac{x_i}{X}} - Y\right)^2.$$

Using (9.5.11) and (9.5.12), an unbiased estimator of the variance of $\hat{\hat{Y}}$ is obtained as

$$\begin{aligned} Est.Var(\hat{\hat{Y}}) = &\left(\frac{1}{n'} - \frac{1}{N}\right)\frac{1}{n(n'-1)}\left[x'\sum_{i=1}^{n}\frac{y_i^2}{x_i} - \frac{x'^2(A-B)}{n'(n-1)}\right] \\ &+ \frac{1}{n(n-1)}\sum_{i=1}^{n}\left(\frac{x'y_i}{n'x_i} - \hat{\hat{Y}}\right)^2. \end{aligned} \qquad (9.5.13)$$

Thus, the theorem is proved.

9.6 Double Sampling for Unbiased Ratio Estimation

The initial sample is a SRSWOR of size n' selected from a population of N. The second sample is a subsample of size n selected from the initial sample with *pp* to aggregate x, a variable measured in the first sample. Then, the following estimator of $\bar{Y}$ is proposed

$$\overset{*}{\bar{Y}} = \frac{\bar{y}_n}{\bar{x}_n}\bar{x}_{n'}$$

where $\bar{y}_n$ and $\bar{x}_n$ are the sample means of y and x in the second sample and $\bar{x}_{n'}$ is the sample mean of x in the first sample.

We now establish the properties of $\overset{*}{\bar{Y}}$ in the following theorem.

Theorem 9.6.1.

(i) $\overset{*}{\bar{Y}} = \frac{\bar{y}_n}{\bar{x}_n}\bar{x}_{n'}$ *is an unbiased estimator of the population mean* $\bar{Y}$.

(ii) $Var(\overset{*}{\bar{Y}}) = \left[n\binom{n'}{n}\binom{N}{n'}\right]^{-1} \sum'' \bar{x}_{n'} \left(\sum' \frac{\sum_{i=1}^n y_i^2}{\sum_{i=1}^n x_i^2}\right) - \bar{Y}^2$ *where* $\sum'$ *denotes summation over all possible samples of size* n *drawn from the first sample, and* $\sum''$ *denotes summation over all possible samples of size* n' *drawn with SRSWOR from a population of* N *units.*

(iii) *An unbiased estimator of* $Var(\overset{*}{\bar{Y}})$ *is given by*

$$Est.Var(\overset{*}{\bar{Y}}) = \overset{*}{\bar{Y}}{}^2 - G,$$

where

$$\begin{aligned} G &= \frac{\bar{x}_{n'}}{Nn\bar{x}_n}\left[\sum_{i=1}^n y_i^2 + \frac{(N-1)}{(n-1)}\sum_{i\neq j}^n y_i y_j\right] \\ &= \frac{\bar{x}_{n'}}{Nn\bar{x}_n}[Nn\bar{y}_n^2 - (N-n)s_y^2]. \end{aligned}$$

Proof. Given the first sample, the probability of obtaining the second sample is proportion to $\sum_{i=1}^n x_i$. Thus if $P(s)$ denotes the probability of selecting the second sample, then

$$P(s) \propto \sum_{i=1}^n x.$$

We then get, summing over all possible samples

$$1 = \sum_s P(s) = \lambda \cdot \sum_s \left(\sum_{i=1}^{n} x \right)$$
$$= \lambda \cdot \binom{n'-1}{n-1} x',$$

from which one obtains

$$\lambda = \frac{1}{\binom{n'-1}{n-1} x'}.$$

Hence,

$$P(s) = \frac{x}{\binom{n'-1}{n-1} x'} \tag{9.6.1}$$

where x and x' denote the totals of x in the second and first samples. Then, given the first sample,

$$\begin{aligned}
E_2(\overset{*}{\overline{Y}}) &= E_2\left(\frac{\bar{y}_n}{\bar{x}_n} \bar{x}_{n'} \right) \\
&= E_2 \left[\frac{1}{n} \frac{\sum_{i=1}^{n} y_i}{\bar{x}_n} \cdot \bar{x}_{n'} \right] \\
&= \sum_s \left[\frac{1}{n} \frac{\sum_{i=1}^{n} y}{\bar{x}_n} \cdot \bar{x}_{n'} \cdot \frac{x}{\binom{n'-1}{n-1} x'} \right] \\
&= \frac{1}{\binom{n'-1}{n-1} n'} \sum_s \left(\sum_{i=1}^{n} y \right) \\
&= \frac{1}{\binom{n'-1}{n-1} n'} \cdot \binom{n'-1}{n-1} \cdot y' \\
&= \bar{y}_{n'}.
\end{aligned} \tag{9.6.2}$$

Hence, we get

$$E(\overset{*}{\overline{Y}}) = E_1 E_2(\overset{*}{\bar{Y}})$$
$$= E_1(\bar{y}_{n'}) = \overline{Y}, \qquad (9.6.3)$$

which proves part (i) of the theorem. Now, given the first sample,

$$E_2(\overset{*}{\overline{Y}})^2 = E_2\left[\frac{\bar{y}_n^2}{\bar{x}_n^2}\cdot \bar{x}_{n'}^2\right]$$
$$= \sum_s \frac{\bar{y}_n^2}{\bar{x}_n^2}\cdot \bar{x}_{n'}^2 \cdot P(s)$$
$$= \Sigma' \frac{\bar{y}_n^2}{\bar{x}_n^2}\cdot \bar{x}_{n'}^2 \frac{x}{\binom{n'-1}{n-1}x'}$$
$$= \Sigma' \frac{\bar{y}_n^2}{\bar{x}_n}\cdot \bar{x}_{n'} \cdot \frac{1}{\binom{n'}{n}} \qquad (9.6.4)$$

where Σ' denotes summation over all possible samples of size n drawn from the first sample. Then, we get

$$E_1 E_2(\overset{*}{\overline{Y}}) = \Sigma'' \cdot \frac{1}{\binom{N}{n'}}\left[\Sigma' \frac{\bar{y}_n^2}{\bar{x}_n}\cdot \bar{x}_{n'} \cdot \frac{1}{\binom{n'}{n}}\right]$$
$$= \frac{1}{\binom{N}{n'}}\frac{1}{\binom{n'}{n}}\Sigma'' \bar{x}_{n'} \cdot \left\{\frac{1}{n}\Sigma' \frac{y^2}{x}\right\}$$
$$= \left[n\binom{N}{n'}\frac{1}{\binom{n'}{n}}\right]^{-1} \cdot \Sigma'' \bar{x}_{n'}\left(\Sigma' \frac{y^2}{x}\right), \qquad (9.6.5)$$

where Σ'' denotes summation over all possible samples of size n' drawn with SRSWOR from a population of N units, y and x denote totals of y and x in the second sample. Now,

$$Var(\overset{*}{\overline{Y}}) = E(\overset{*}{\overline{Y}})^2 - E^2(\overset{*}{\overline{Y}})$$
$$= n\left[\binom{N}{n'}\frac{1}{\binom{n'}{n}}\right]^{-1} \Sigma''' \bar{x}_{n'}\left(\frac{\Sigma' y^2}{x}\right) - \overline{Y}^2. \qquad (9.6.6)$$

Thus, part (ii) of the theorem is proved.

Now, we consider the derivation of estimator of $Var(\overset{*}{\overline{Y}})$. We have

$$E\left[\frac{\sum_{i=1}^{n} y_i^2}{\sum_{i=1}^{n} x_i}\right] = E_1 E_2 \left[\frac{\sum_{i=1}^{n} y_i^2}{n\bar{x}} \cdot \frac{\bar{x}_{n'}}{n'}\right]$$

$$= E_1 \left[\Sigma' \left\{\frac{\sum_{i=1}^{n} y_i^2}{n\bar{x}} \cdot \bar{x}_{n'} \cdot P(S)\right\}\right]$$

$$= E_1 \left[\Sigma' \left\{\frac{\sum_{i=1}^{n} y_i^2}{n\bar{x}} \cdot x_{n'} \cdot \frac{n\bar{x}_n}{\binom{n'-1}{n-1} n'\bar{x}_n}\right\}\right]$$

$$= E_1 \left[\Sigma' \frac{1}{n'\binom{n'-1}{n-1}} \sum_{i=1}^{n} y_i^2\right]$$

$$= E_1 \left[\frac{1}{n'\binom{n'-1}{n-1}} \cdot \binom{n'-1}{n-1} \sum_{i=1}^{n'} y_i^2\right]$$

$$= E_1 \left[\frac{1}{n'} \sum_{i=1}^{n'} y_i^2\right] = \frac{1}{N} \sum_{i=1}^{N} y_i^2. \tag{9.6.7}$$

Further,

$$E\left[\frac{\sum_{i \neq j}^{n} y_i y_j}{\sum_{i=1}^{n} x_i} \cdot \bar{x}_{n'}\right] = E_1 E_2 \left[\frac{\sum_{i \neq j}^{n} y_i y_j}{n\bar{x}_n} \cdot \frac{\bar{x}_{n'}}{n'}\right]$$

$$= E_1 \left[\Sigma' \left\{\frac{\sum_{i \neq j}^{n} y_i y_j}{n\bar{x}_n} \cdot \bar{x}_{n'} \cdot P(s)\right\}\right]$$

$$= E_1 \left[\Sigma' \left\{ \frac{\sum_{i \neq j}^{n} y_i y_j}{n\bar{x}_n} \cdot x_{n'} \cdot \frac{n\bar{x}_n}{\binom{n'-1}{n-1} n\bar{x}_{n'}} \right\} \right]$$

$$= E_1 \left[\Sigma' \frac{1}{n' \binom{n'-1}{n-1}} \sum_{i \neq j}^{n} y_i y_j \right]$$

$$= E_1 \left[\frac{1}{n' \binom{n'-1}{n-1}} \binom{n'-2}{n-2} \sum_{i \neq j}^{n'} y_i y_j \right]$$

$$= E_1 \left[\frac{(n-1)}{n'(n'-1)} \cdot \sum_{i \neq j}^{n'} y_i y_j \right]$$

$$= \frac{(n-1)}{n'(n'-1)} \cdot \frac{n'(n'-1)}{N(N-1)} \sum_{i \neq j}^{N} y_i y_j$$

$$= \frac{(n-1)}{N(N-1)} \sum_{i \neq j}^{N} y_i y_j. \qquad (9.6.8)$$

From (9.6.7) and (9.6.8), we obtain

$$E \left[\frac{1}{N} \frac{\sum_{i=1}^{n} y_i^2}{\sum_{i=1}^{n} x_i} \cdot \bar{x}_{n'} + \frac{(N-1)}{N(n-1)} \frac{\sum_{i \neq j}^{n} y_i y_j}{\sum_{i=1}^{n} x_i} \bar{x}_{n'} \right] = \frac{1}{N^2} \left[\sum_{i=1}^{N} y_i^2 + \sum_{i \neq j}^{N} y_i y_j \right]$$

$$= \bar{Y}^2. \qquad (9.6.9)$$

From (9.6.5), we see that

$$E(\overset{*}{\bar{Y}})^2 = \left[n \binom{N}{n'} \binom{n'}{n} \right]^{-1} \Sigma'' \bar{x}_{n'} \left(\Sigma' \frac{y^2}{x} \right). \qquad (9.6.10)$$

Finally (9.6.9) and (9.6.10), we obtain an unbiased estimator of $Var(\overset{*}{\bar{Y}})$ as

$$Est.Var(\overset{*}{\bar{Y}}) = \overset{*}{\bar{Y}}^2 - G, \qquad (9.6.11)$$

where

$$G = \frac{\bar{x}_{n'}}{N}\left[\frac{\sum_{i=1}^{n} y_i^2}{\sum_{i=1}^{n} x_i} + \frac{(N-1)}{(n-1)}\frac{\sum_{i\neq j}^{n} y_i y_j}{\sum_{i=1}^{n} x_i}\right]$$

$$= \frac{\bar{x}_{n'}}{Nn\bar{x}_n}\left[\sum_{i=1}^{n} y_i^2 + \frac{(N-1)}{(n-1)}\sum_{i\neq j}^{n} y_i y_j\right]$$

$$= \frac{\bar{x}_{n'}}{Nn\bar{x}_n}[nN\bar{y}_n^2 - (N-n)s_y^2].$$

9.7 Successive Sampling

In the previous chapters, we have considered the theory of surveys which are made on one occasion. But sometimes surveys are repeated over the same population regularly over a period of time. When sampling is done over the same population on more than one occasion, we call it successive sampilng. The advantage of successive sampling is that the sampler is in an ideal position to make realistic estimates both of costs and of variances and to apply the techniques that lead to optimum efficiency. Many government surveys are of this type. The reason for discussing repetitive surveys or successive sampling is that they have some similarities to double sampling procedures. For example, a first sample is taken on one occasion and a second sample is to be taken on another occasion.

In successive sampling, we wish to estimate the following three quantities:

(i) The change in the population mean $\bar{Y}$ from one occasion to the next.
(ii) The average value of $\bar{Y}$ over all occasions.
(iii) The average value $\bar{Y}$ for the current occasion.

In most surveys, the main purpose is to estimate (iii), the current $\bar{Y}$.

We shall illustrate the technique of successive sampling by considering sampling on two occasions.

We suppose that the samples are of the same size n on both occasions, and we wish to estimate the current average $\bar{Y}$ of the population. We assume that simple random sampling is used and that fpc may be ignored.

Table 9.2 Estimators from the unmatched and the matched portion.

	Estimator	Variance
Unmatched	$\bar{y}'_{2u} = \bar{y}_{2u}$	$\frac{S_2^2}{u} = \frac{1}{W_{2u}}$
Matched	$\bar{y}'_{2m} = y + b(\bar{x}_m - \bar{x}_n)$	$\frac{S_2^2(1-\rho^2)}{m} + \frac{\rho^2 S_2^2}{n} = \frac{1}{W_{2m}}$

On the first occasion, a SRS of size n is taken. A random subsample of m units in the first sample is retained (m for matched for use on a second occasion and the remaining u units (u for unmatched) are discarded and are replaced by a new selection of a second sample from the units of the population not previously selected. We use the following notations.

$\bar{y}_{2u}$ = mean of unmatched portion on the second occasion.

$\bar{y}_{2m}$ = mean of the matched portion on the second occasion.

$\bar{y}_2$ = mean of the whole sample on the second occasion.

$\bar{Y}_2$ = the population mean on the second occasion.

S_2^2 = the population mean sum of square on the second occasion.

The matched and unmatched portions of the second sample provide independent estimators y'_{2m}, y'_{2u} of Y as shown in Table 9.2. In the matched portion, we use double sampling regression estimate, where the large sample is the first sample and the auxiliary variate x_i is the value of y_i on the first occasion.

In the double-sampling regression estimator $\bar{y}'_{2m}$, $\bar{x}_n$ is the mean of the first sample and $\bar{x}_m$ denotes the mean of the matched portion on the first occasion. The variance of $\bar{y}'_{2m}$ is obtained from (9.4.3).

The best combined estimator of $\bar{Y}_2$ is obtained by weighting the two independent estimators $\bar{y}'_{2u}$ and $\bar{y}'_{2m}$ inversely as their variances. Let W_{2u}, and W_{2m} be the inverses of the variances of $\bar{y}'_{2u}$ and $\bar{y}'_{2m}$ respectively. Then, using least squares theory, the best combined estimators of $\bar{Y}_2$ is obtained as

$$\bar{y}'_2 = \lambda \bar{y}'_{2u} + (1 - \lambda)\bar{y}'_{2m}, \tag{9.7.1}$$

where

$$\lambda = \frac{W_{2u}}{W_{2u} + W_{2m}}$$

and the variance of $\bar{y}'_2$ is given by

$$Var(\bar{y}'_2) = \frac{1}{W_{2u} + W_{2m}}. \tag{9.7.2}$$

Substituting the values of W_{2u} and W_{2m} from Table 9.2, in (9.7.2), we obtain the variance of $\bar{y}_2'$ as

$$Var(\bar{y}_2') = \frac{S_2^2(n - u\rho^2)}{n^2 - u^2\rho^2}. \tag{9.7.3}$$

It may be noted here that if $u = 0$, or $u = n$ (no matching), then (9.7.3) has the same value $\frac{S_2^2}{n}$. We can find the optimum value of u by minimizing the variance of $\bar{y}_2'$. Equating the derivative of (9.7.3) with respect to u to zero, we obtain

$$u^2\rho^2 - 2nu + n^2 = 0,$$

from which we obtain the optimum value of u as

$$u = n\left[\frac{1 \pm \sqrt{1-\rho^2}}{2}\right].$$

Since u is less than n, we take the optimum value of u, by considering the negative sign in the roots, and obtain

$$u = n\left[\frac{1 - \sqrt{1-\rho^2}}{\rho^2}\right] = \frac{n}{1 + \sqrt{1-\rho^2}}. \tag{9.7.4}$$

Hence, the optimum value of m is obtained as

$$m = n - u = \frac{n\sqrt{1-\rho^2}}{[1 + \sqrt{1-\rho^2}]}. \tag{9.7.5}$$

The minimum value of variance of $\bar{y}_2'$ is obtained by substituting the optimum value of u in (9.7.3) and is given by

$$Var(\bar{y}_2')_{min} = \frac{S_2^2}{2n}(1 + \sqrt{1-\rho^2}). \tag{9.7.6}$$

The method can be generalized to sampling on more than two occasions. For this, we refer to Yates (1960) and Patterson (1950).

EXERCISES

9.1. Consider the double sampling in which the first sample is a SRSWOR of size n' drawn from a population of size N and information on x is obtained. Another independent sample of size n is selected with *pp* to x according to Lahiri's method. Show that

(i) $\hat{Y} = \left(\frac{N}{n'}x'\right)\left(\frac{1}{n}\sum_{i=1}^{n}\frac{y_i}{x_i}\right)$ is an unbiased estimator of the population total Y.

(ii) $Var(\hat{Y}) = \left(\frac{1}{n'} - \frac{1}{N}\right)N^2R^2S_x^2 + \frac{1}{n}V_p(y)\left[1 + \left(\frac{1}{n'} - \frac{1}{N}\right)\frac{S_x^2}{\bar{X}^2}\right]$.

(iii) An unbiased estimator of $Var(\hat{Y})$ is given by

$$Est.Var(\hat{Y}) = \left(\frac{1}{n}\sum_{i=1}^{n}\frac{y_i}{x_i}\right)^2 N^2\left(\frac{1}{n'} - \frac{1}{N}\right)s_x^2$$
$$+\frac{N^2}{n(n-1)n'}\sum_{i=1}^{n}\left(\frac{y_i}{x_i} - \hat{R}\right)^2\left[\frac{x'^2}{n'} - \left(1 - \frac{n'}{N}\right)s_x^2\right],$$

where $R = \frac{Y}{X}$, $\hat{R} = \frac{1}{n}\sum_{i=1}^{n}\left(\frac{y_i}{x_i}\right)$,

$$V_p(y) = \sum_{i=1}^{N}\frac{x_i}{X}\left(\frac{y_i}{\frac{x_i}{x}} - Y\right)^2,$$

and s_x^2 = mean sum of squares of x for the second sample.

9.2. \$3000 is allocated for a survey to estimate a proportion. The main survey will cost \$10 per sampling unit. Information is available in files, at a cost of \$0.25 per sampling unit, that enables the units to be classified into two strata of about equal sizes. If the true proportion is 0.2 in stratum 1 and 0.8 in stratum 2, estimate the optimum values of n, n' and the resulting value of $Var(p_{st})$. Does double sampling produce a gain in precision over single sampling? (Ignore the ratios $\frac{n'}{N}$, $\frac{n_i}{N_i}$, $i = 1, 2$.) Also, find the cost ratios $\frac{c_n}{c'_n}$ for which double sampling is more economical than single sampling.

9.3. If $\rho = 0.8$ in double sampling for regression, how large must n' be relative to n, if the loss in precision due to sampling errors in the mean of the large sampe is to be less than 10%?

9.4. An initial sample of n' units is taken from a population of size N with SRSWOR and information on p variates $x_1, x_2, \ldots, x_p$ is collected. From it, a SRSWOR of size n is selected and information on y is collected. Show that an unbiased estimator of the population mean for y is given by $\hat{\bar{Y}} = \sum_{i=1}^{p} w_i u_i$, where $u_i = \bar{y} - k_i(\bar{x}_i - \bar{x}'_i)$, and $\bar{y}$, $\bar{x}_i$, $(i = 1, 2, \ldots, p)$ are the means of the subsample, $\bar{x}'_i$ is the mean of x_i in the initial sample and k_i's are some constants, w_i being the weights adding up to 1. Further, show that $Var(\hat{\hat{Y}}) = n^{-1}\sum_{i=1}^{p}\sum_{j=1}^{p} w_i w_j b_{ij}$, where

$$b_{ij} = \left(1 - \frac{n}{N}\right) S_{oo} + \left(1 - \frac{n}{n'}\right) [k_i k_j S_{ij} - k_i S_{oi} - k_j S_{oj}],$$

and $S_{ij} = Cov(x_i, x_j)$, $S_{oi} = Cov(y, x_i)$, $S_{oo} = S_y^2$. Also, show that an unbiased estimator of the $Var(\hat{\hat{Y}})$ is given by

$$\left(\frac{1}{n'} - \frac{1}{N}\right)(n-1)^{-1}\sum_{i=1}^{n}(y_i - \bar{y})^2 + \left(\frac{1}{n} - \frac{1}{n'}\right)(n-1)^{-1}$$

$$\cdot \sum_{i=1}^{n}\left[y_i - \bar{y} - \sum_{j=1}^{p} w_j k_j (x_{ji} - \bar{x}_j)\right]^2 .$$

9.5. A SRSWOR of n' units is selected and information on x variate is collected. From it a subsample of size n is selected with SRSWOR and y is measured for it. The following two estimators for the population $\bar{Y}$ are suggested.

(i) $\hat{\bar{Y}}_1 = \bar{x}'_{n'} \frac{\bar{y}_n}{\bar{x}_n}$.

(ii) $\hat{\bar{Y}}_2 = \bar{r}_n \bar{x}_{n'} + \frac{n(n'-1)}{n'(n-1)}(\bar{y}_n - \bar{r}_n \bar{x}_n)$, where $\bar{r}_n = \frac{1}{n}\sum_{i=1}^{n}\left(\frac{y_i}{x_i}\right)$.

Calculate $Var(\hat{\bar{Y}}_1)$ and $Var(\hat{\bar{Y}}_2)$ and obtain conditions under which $\hat{\bar{Y}}_2$ is more efficient than $\hat{\bar{Y}}_1$.

9.6. Consider the estimator

$$\bar{y}_d = \bar{y}_n + b(\bar{x}_{n'} - \bar{x}_n),$$

where n is a subsample of n' and (y_i, x_i) follows a bivariate normal distribution with means (μ_y, μ_x), variances (σ_y^2, σ_x^2) and correlation coefficient ρ. Show that $\bar{y}_d$ is an unbiased estimator of μ_y and that its variance is given by

$$Var(\bar{y}_d) = \sigma_y^2(1-\rho^2)\left(\frac{1}{n} - \frac{1}{n'}\right)\left[1 + \frac{1}{(n-3)}\right] + \frac{\sigma_y^2}{n'}.$$

Further, show that an unbiased estimator of $Var(\bar{y}_d)$ is given by

$$Est.Var(\bar{y}_d) = \left(\frac{1}{n} - \frac{1}{n'}\right)\frac{T}{(n-3)} + \frac{s_y^2}{n'},$$

where $T = \sum_{i=1}^{n}[(y_i - \bar{y}_n) - b(x_i - \bar{x}_n)]^2$.

9.7. In simple random sampling on two occasions, let the estimator of the mean $\bar{Y}_2$ on the second occasion, in the notation of Section 9.7, be defined as

$$\bar{y}_2'' = (1-\lambda)(\bar{y}_{2m} + \bar{x}_n - \bar{x}_m) + \lambda \bar{y}_{2u}.$$

(i) Ignoring fpc, show that

$$Var(\bar{y}_2'' = \frac{S_2^2}{n}\left[(1-\lambda)^2 \frac{[1+(1-2\rho)]}{\phi} + \frac{\lambda^2}{\mu}\right],$$

where $\phi = \frac{m}{n}$, $\mu = \frac{u}{n}$.

(ii) For given ρ, ϕ, μ, find the value of λ which minimizes $Var(\bar{y}_2'')$. Show that if $\rho > \frac{1}{2}$, then the best weight λ lies between μ and $\frac{\mu}{(\mu+1)}$.

9.8. In sampling on two occasions, suppose that $S_1 = S_2 = S$ and that the samples are large, so that the regression coefficients of y on x and of x on y in the matched part of the samples on the two occasions are both effectively equal to ρ. The estimator $\bar{y}_2'$ given by (9.7.1) is constructed and an analogous estimator using regression of x on y is also constructed. Show that

$$Var(\bar{y}_2' - \bar{y}_1') = \frac{2S^2(1-\rho)}{(n-u\rho)}$$

$$Var(\bar{y}_2' + \bar{y}_1') = \frac{2S^2(1+\rho)}{(n+u\rho)}.$$

9.9. A survey is to be planned on two occasions to estimate $\theta = a\bar{Y}_1 + b\bar{Y}_2$, where $\bar{Y}_1$, $\bar{Y}_2$ are the means on two occasions anad a and b are known constants. A sample of n units will be taken on the first occasion. On the second occasion a subsample of n_1 units will be taken, as well as an independent sample of $n_2 = n - n_1$ units. If the total cost of the survey is given by the function $c = c_0 + cn + c_1 n_1 + c_2 n_2$, obtain the best values of n and n_1 so that for a specified cost c_0, the variance of the estimator is minimized.

Chapter 10

NON-SAMPLING ERRORS

10.1 Non-sampling Errors

In the sampling theory developed in the previous chapters, it was assumed that the true value of each unit in the population can be obtained and tabulated without errors. But this is not so in practice. In practice, there are errors of observations and errors in tabulation. Such errors which are due to factors other than sampling are called non-sampling errors. In large-scale censuses and surveys, non-sampling errors are unavoidable. Thus, data collected in a census by complete enumeration, although free from sampling error, would not be free from non-sampling error. The data collected in a sample survey would be subjected to sampling error as well as non-sampling error. Generally, sampling error decreases with increase in sample size, while non-sampling error increases with increase in the sample size.

In recent years, need for assessing and controlling the non-sampling errors at various stages of collection and tabluation of data in large-scale censuses and surveys has been greatly felt. In this chapter, we shall consider most of the sources and types of non-sampling errors and study the techniques for assessment and control of these errors.

In survey work, it is assumed that the value of the characteristic to be measured has been precisely defined for every population unit and for any given population unit, this value exists and is unique. This value is called the true value of the characteristic for the population value. In practice, data collected on the selected units are called survey values and are different from the true values. This discrepancy between the true value and survey value is called the observational error or response error, and arises primarily from the variable performance of the investigators and lack of precision in measurement techniques.

10.2 Sources of Non-sampling Errors

The main sources of non-sampling errors are lack of precision in measurement techniques, lack of proper specification of domain of study and scope of investigation, incomplete coverage of the population or sample, defective methods of collection of data, tabulation errors and variable performance of investigators. Non-sampling errors may arise from one or more of the following factors:

(i) Omission or duplication of units due to imprecise definition of the boundaries of area units, incomplete or wrong identification particulars of units or faulty methods of enumeration.
(ii) Inaccurate or inappropriate methods of interview.
(iii) Observations or measurements with inadequat or ambiguous questionnaires, definitions or instructions.
(iv) Lack of trained and experienced investigators.
(v) Lack of adequate inspection and supervision of investigators.
(vi) Errors in data processing, such as coding, punching, verification, tabulation, etc.
(vii) Errors in printing of tabulated results.
(viii) Biases of the investigators.

The above sources are not exhaustive. Non-sampling errors may be broadly classified into three types: (i) specification errors, (ii) observational errors, and (iii) tabulation errors. Specification errors occur at the planning stage. Observational errors occur at the field work of the survey and tabulation errors occur at the stage of tabulation or processing of data. In the next section, we shall consider the treatment of observational errors.

10.3 Observational Errors

Let x_i denote the true value of the characteristic on the ith unit in SRSWOR of h units drawn from a population of N units, $i = 1, 2, \ldots, h$. Let there be m enumerators forming a simple random sample selected without replacement from a population of M enumerators. We assume that the units in the sample of h are randomly allotted to the different enumerators. We also assume that (i) each enumerator makes an equal number of observations $\bar{n}$, say and (ii) the number of observations made on any unit in the sample is equal for all units, and is given by p, say. Let

n_{ij} = the number of observations made on the ith unit by the jth enumerator (1 or 0).

$$n_{i\cdot} = \sum_{j=1}^{m} n_{ij} = \text{the number of observations made on the } i^{th} \text{ unit}$$

$$= p, \text{ for all } i = 1, 2, \ldots, h.$$

$$n_{\cdot j} = \sum_{i=1}^{h} n_{ij} = \text{the numer of observations made by the } j^{th} \text{ enumerator}$$

$$= \bar{n}, \text{ for } j = 1, 2, \ldots, m.$$

$$n = \sum_{i=1}^{h}\sum_{j=1}^{m} n_{ij} = \text{total number of observations made}$$

$$= m\bar{n} = ph.$$

y_{ij} = the value reported by the jth enumerator on the ith unit.

$\bar{y}_{\cdot j}$ = the mean of all $\bar{n}$ observations made by the jth enumerator

$$= \frac{\sum_{i}^{n} y_{ij}}{\bar{n}}.$$

$\bar{y}_{\cdot\cdot}$ = the mean of all n observations made on all the h units in the sample

$$= \frac{\sum_{i=1}^{h}\sum_{j=1}^{m} y_{ij}}{n} = \frac{1}{m}\sum_{j=1}^{m} \bar{y}_{\cdot j}.$$

We assume the following mathematical model.

$$y_{ij} = x_i + \alpha_j + \epsilon_{ij}, \quad i = 1, \ldots, h;\ j = 1, \ldots, m, \tag{10.3.1}$$

where α_j represents the bias of the jth enumerator and ϵ's are random errors distributed independently with mean 0 and a common variance S_ϵ^2.

From (10.3.1), we obtain

$$\bar{y}_{\cdot j} = \frac{1}{\bar{n}}\sum_{i=1}^{h} x_i n_{ij} + \alpha_j + \frac{1}{\bar{n}}\sum_{i=1}^{h} \epsilon_{ij} n_{ij}. \tag{10.3.2}$$

Summing (10.3.2) over all j and dividing the sum by m, we obtain

$$\bar{y}_{..} = \frac{1}{h}\sum_{i=1}^{h} x_i + \frac{1}{m}\sum_{j=1}^{m}\alpha_j + \frac{1}{n}\sum_{i=1}^{n}\sum_{j=1}^{m}\epsilon_{ij}n_{ij}. \qquad (10.3.3)$$

Taking expectation of (10.3.2), we get

$$E(\bar{y}_{.j}) = E\left[\frac{1}{\bar{n}}\sum_{i=1}^{h} x_i n_{ij}\right] + \frac{1}{M}\sum_{j=1}^{M}\alpha_j. \qquad (10.3.4)$$

Now, $\frac{1}{\bar{n}}\sum_{i=1}^{h} x_i n_{ij}$ is the sample mean of $\bar{n}$ units assigned to the jth enumerator. Since, these $\bar{n}$ units are selected at random out of the h units, they constitute a SRSWOR of $\bar{n}$ units selected from h units. Hence

$$\begin{aligned} E\left[\frac{1}{\bar{n}}\sum_{i=1}^{h} x_i n_{ij}\right] &= E_1\left[E_2\left(\frac{1}{\bar{n}}\sum_{i=1}^{h} x_i n_{ij}|h\right)\right] \\ &= E_1\left[\frac{1}{n}\sum_{i=1}^{h} x_i\right] \\ &= \bar{X}, \end{aligned} \qquad (10.3.5)$$

where $\bar{X}$ is the true population mean. Using (10.3.5) in (10.3.4), we get

$$E(\bar{y}_{.j}) = \bar{X} + \bar{\alpha}. \qquad (10.3.6)$$

Also, from (10.3.3), one can easily obtain

$$E(\bar{y}_{..}) = \bar{X} + \bar{\alpha}. \qquad (10.3.7)$$

From (10.3.7), we see that the observed sample mean $\bar{y}_{..}$ is not an unbiased estimator of the true mean. However, if the enumerator's biases vary in such a way that $\bar{\alpha} = 0$, then $\bar{y}_{..}$ will be an unbiased estimtor of $\bar{X}$. But in practice $\bar{\alpha} \neq 0$ where $\bar{\alpha}$ is the non-sampling bias. In sample surveys $\bar{\alpha}$ must be kept as small as possible.

In the following theorem, we derive the variance of the sample mean $\bar{y}_{..}$.

Theorem 10.3.1. *The variance of $\bar{y}_{..}$ is given by*

$$Var(\bar{y}_{..}) = \left(\frac{1}{h} - \frac{1}{N}\right)S_x^2 + \left(\frac{1}{m} - \frac{1}{M}\right)S_\alpha^2 + \frac{1}{n}S_\epsilon^2$$

where

$$S_x^2 = \frac{1}{(N-1)} \sum_{i=1}^{N} (x_i - \bar{X})^2$$

$$S_\alpha^2 = \frac{1}{M-1} \sum_{j=1}^{M} (\alpha_j - \bar{\alpha})^2.$$

Proof. The variance of $\bar{y}..$ is given by

$$\begin{aligned} Var(\bar{y}..) &= E[\bar{y}.. - \bar{X} - \bar{\alpha}]^2 \\ &= E\left[\left(\frac{1}{h}\sum_{i=1}^{h} x_i - \bar{X}\right) + \left(\frac{1}{m}\sum_{j=1}^{m} \alpha_j - \bar{\alpha}\right) + \frac{1}{n}\sum_{i=1}^{h}\sum_{j=1}^{m} n_{ij}\epsilon_{ij}\right]^2 \\ &= E\left[\frac{1}{h}\sum_{i=1}^{h} x_i - \bar{X}\right]^2 + E\left[\frac{1}{m}\sum_{j=1}^{m} \alpha_j - \bar{\alpha}\right]^2 \\ &\quad + E\left[\frac{1}{n}\sum_{i=1}^{h}\sum_{j=1}^{m} n_{ij}\epsilon_{ij}\right]^2. \end{aligned} \qquad (10.3.8)$$

Now, $\frac{1}{h}\sum_{i=1}^{h} x_i$ is the mean of a simple random sample of size h drawn without replacement from the population of N units. Hence

$$E\left[\frac{1}{h}\sum_{i=1}^{h} x_i - \bar{X}\right]^2 = \left(\frac{1}{h} - \frac{1}{N}\right) S_x^2. \qquad (10.3.9)$$

Similarly, we have

$$E\left[\frac{1}{m}\sum_{j=1}^{m} \alpha_j - \bar{\alpha}\right] = \left(\frac{1}{m} - \frac{1}{M}\right) S_\alpha^2. \qquad (10.3.10)$$

Further, $\frac{1}{n}\sum_{i=1}^{h}\sum_{j=1}^{m} n_{ij}\epsilon_{ij}$ is the mean of n errors corresponding to n observations made by all the enumerators, hence

$$E\left[\frac{1}{n}\sum_{i=1}^{h}\sum_{j=1}^{m} n_{ij}\epsilon_{ij}\right]^2 = \frac{1}{n} S_\epsilon^2. \qquad (10.3.11)$$

Using (10.3.9), (10.3.10) and (10.3.11) in (10.3.8), we obtain

$$Var(\bar{y}..) = \left(\frac{1}{h} - \frac{1}{N}\right) S_x^2 + \left(\frac{1}{m} - \frac{1}{M}\right) S_\alpha^2 + \frac{1}{n} S_\epsilon^2. \qquad (10.3.12)$$

If N and M are large, then we obtain from (10.3.12)

$$Var(\bar{y}_{..}) = \frac{S_x^2}{h} + \frac{S_\alpha^2}{m} + \frac{S_\epsilon^2}{n}. \tag{10.3.13}$$

In practice, usually $p = 1$, that is, one observation is made on each unit. Then, $n = hp = h$ and hence, we get

$$\begin{aligned} Var(\bar{y}_{..}) &= \frac{S_x^2 + S_\epsilon^2}{h} + \frac{S_\alpha^2}{m} \\ &= \frac{S_x^2 + S_\alpha^2 + S_\epsilon^2}{h} + \left(\frac{1}{m} - \frac{1}{h}\right) S_\alpha^2 \\ &= \frac{S_y^2}{h} + \left(\frac{1}{m} - \frac{1}{h}\right) S_\alpha^2. \end{aligned} \tag{10.3.14}$$

We note that m is usually less than h and hence the second term in (10.3.14) is positive, unless S_α^2 is zero, that is, the variation in the biases of the enumerators is zero. We thus see that the sampling variance of the sample mean is not entirely due to errors arising from chance variations in the selection of the sample of h units, but is inflated by the variability in the biases of the enumerators.

There is another alternative expression for the variance of $\bar{y}_{..}$ due to Hansen et al. (1953). This expression is in terms of the correlation coefficient ρ between the responses obtained by the same numerator. By definition of ρ, we have

$$\begin{aligned} \rho S_y^2 \left(1 - \frac{1}{N}\right) &= E[\{y_{ij} - E(y_{ij})\}\{y_{kj} - E(y_{kj}\}] \\ &= E_1\{E_2[\{y_{ij} - E(y_{ij})\}\{y_{kj} - E(y_{kj})\}|j]\} \\ &= E_1\{E_2[\{x_i - X + \alpha_j - \bar{\alpha} + \epsilon_{ij}\} \\ &\quad \cdot \{x_k - \bar{X} + \alpha_j - \bar{\alpha} + \epsilon_{kj}\}|j]\} \\ &= E[(x_i - \bar{X})(x_k - \bar{X})] + E[(\alpha_j - \bar{\alpha})^2], \end{aligned} \tag{10.3.15}$$

since the expectations of other terms are zero. From (10.3.15), we get

$$\rho S_y^2 \left(1 - \frac{1}{N}\right) = -\frac{S_x^2}{N} + \left(1 - \frac{1}{M}\right) S_\alpha^2. \tag{10.3.16}$$

If N and M are large, (10.3.16) gives

$$\rho S_y^2 = S_\alpha^2, \tag{10.3.17}$$

which on substitution in (10.3.14) gives

$$Var(\bar{y}_{..}) = \frac{S_y^2}{h}\left[1 + \rho\left(\frac{h}{m} - 1\right)\right]. \tag{10.3.18}$$

We also note that the variance of the sample mean $\bar{y}_{..}$ does not reduce to zero even when a complete count is made. For, consider the limit of (10.3.14) when $N \to \infty$. When $N \to \infty$, suppose m' enumerators are required and selected from a population of M' enumerators, then h is also very large. Let S'^2_α represent the variability of the biases of the population of the enumerators. Then, from (10.3.14), we obtain, when N,

$$Var(\bar{y}_{..}) = \frac{1}{m'} S'^2_\alpha.$$

We see that a sample may give a more efficient estimator than a complete census if

$$\frac{S'^2_\alpha}{m'} > \frac{S^2_y}{h} + \left(\frac{1}{m} - \frac{1}{h}\right) S^2_\alpha.$$

10.4 Estimation of the Variance

We shall show how to obtain estimators of the different components of the variance of the sample mean $\bar{y}_{...}$.

First, we shall derive the variace of $\bar{y}_{.j}$, the sample mean for the jth enumerator.

$$\begin{aligned} Var(\bar{y}_{.j}) &= E[\bar{y}_{.j} - E(\bar{y}_{.j})]^2 \\ &= E\left[\left(\frac{1}{\bar{n}}\sum_{i=1}^{h} x_i n_{ij} - \bar{X}\right) + (\alpha_j - \bar{\alpha}) + \frac{1}{\bar{n}}\sum_{i=1}^{h} n_{ij}\epsilon_{ij}\right]^2 \\ &= E\left[\left(\frac{1}{\bar{n}}\sum_{i=1}^{h} x_i n_{ij} - \bar{X}\right)^2\right] + E(\alpha_j - \bar{\alpha})^2 + E\left[\frac{1}{\bar{n}}\sum_{i=1}^{h} \epsilon_{ij} n_{ij}\right]^2. \end{aligned} \tag{10.4.1}$$

For fixed sample of h units, $\frac{1}{\bar{n}}\sum_{i=1}^{h} x_i n_{ij}$ is the mean of a simple random sample of $\bar{n}$ units selected without replacement from a population of h units, hence

$$E\left[\left(\frac{1}{\bar{n}}\sum_{i=1}^{h} x_i n_{ij} - \bar{X}\right)^2 |h\right] = \left(\frac{1}{\bar{n}} - \frac{1}{h}\right) s^2_x,$$

where $s_x^2 = \frac{1}{h-1}\sum_{i=1}^{h}(x_i - \bar{x})^2$. Hence

$$E\left[\left(\frac{1}{\bar{n}}\sum_{i=1}^{h} x_i n_{ij} - \bar{X}\right)^2\right] = Var\left(\frac{1}{\bar{n}}\sum_{i=1}^{h} x_i n_{ij}\right)$$
$$= V_1\left(E_2\left\{\frac{1}{\bar{n}}\sum_{i=1}^{h} x_i n_{ij}|h\right\}\right)$$
$$+ E_1\left(V_2\left\{\frac{1}{\bar{n}}\sum_{i=1}^{h} x_i n_{ij} h\right\}\right)$$
$$= V_1\left[\frac{1}{h}\sum_{i=1}^{h} x_i\right] + E_1\left[\left(\frac{1}{\bar{n}} - \frac{1}{h}\right) s_x^2\right]$$
$$= \left(\frac{1}{h} - \frac{1}{N}\right) S_x^2 + \left(\frac{1}{\bar{n}} - \frac{1}{h}\right) S_x^2$$
$$= \left(\frac{1}{\bar{n}} - \frac{1}{N}\right) S_x^2. \qquad (10.4.2)$$

Further,

$$E(\alpha_j - \bar{\alpha})^2 = \left(1 - \frac{1}{M}\right) S_\alpha^2. \qquad (10.4.3)$$

Also, $\frac{1}{\bar{n}}\sum_{i=1}^{h} \epsilon_{ij} n_{ij}$ is the mean of $\bar{n}$ errors corresponding to $\bar{n}$ observations made by the jth enumerator. Since ϵ_{ij} have been assumed to be distributed independently with variance S_ϵ^2, it follows that

$$Var\left(\frac{1}{\bar{n}}\sum_{i=1}^{h} \epsilon_{ij} n_{ij}\right) = \frac{S_\epsilon^2}{\bar{n}}. \qquad (10.4.4)$$

Using (10.4.2), (10.4.3) and (10.4.4) in (10.4.1), we get

$$Var(\bar{y}_{\cdot j} = \left(\frac{1}{\bar{n}} - \frac{1}{N}\right) S_x^2 + \left(1 - \frac{1}{M}\right) S_\alpha^2 + \frac{1}{\bar{n}} S_\epsilon^2. \qquad (10.4.5)$$

If N and M are large, then (10.4.5) becomes

$$Var(\bar{y}_{\cdot j}) = \frac{1}{\bar{n}}(S_x^2 + S_\epsilon^2) + S_\alpha^2. \qquad (10.4.6)$$

We shall now consider the problem of estimation of the variance of $\bar{y}_{\cdot\cdot}$. Let s_e^2 denote the mean sum of squares between the means of m enumerations, defined by

$$s_e^2 = \frac{1}{(m-1)}\sum_{j=1}^{m}(\bar{y}_{\cdot j} - \bar{y}_{\cdot\cdot})^2.$$

Then, we have

$$\begin{aligned}(m-1)E(s_e^2) &= E\left[\sum_{j=1}^{m}\bar{y}_{.j}^2 - m\bar{y}_{..}^2\right]\\ &= \sum_{j=1}^{m}E(\bar{y}_{.j}^2) - mE(\bar{y}_{..}^2)\\ &= \sum_{j=1}^{m}\{Var(\bar{y}_{.j} + (E(\bar{y}_{.j}))^2\} - m\{Var(\bar{y}_{..}) + (E(\bar{y}_{..}))^2\}\\ &= \sum_{j=1}^{m}Var(\bar{y}_{.j}) - m\,Var(\bar{y}_{..}). \qquad (10.4.7)\end{aligned}$$

Substitution from (10.4.6) and (10.3.13) in (10.4.7) and noting that $n = ph$, we obtain

$$E(s_e^2) = \frac{m(m-p)}{hp(m-1)}S_x^2 + S_\alpha^2 + \frac{m}{np}S_\epsilon^2. \qquad (10.4.8)$$

Next, let s_w^2 denote the mean sum of squares between observations within enumerators, defined by

$$\begin{aligned}s_w^2 &= \frac{1}{m(\bar{n}-1)}\sum_{j=1}^{m}\sum_{i=1}^{\bar{n}}(y_{ij} - \bar{y}_{.j})^2\\ &= \frac{1}{(n-m)}\sum_{j=1}^{m}\sum_{i=1}^{\bar{n}}(y_{ij} - \bar{y}_{.j})^2. \qquad (10.4.9)\end{aligned}$$

Then, we obtain

$$\begin{aligned}(n-m)E(s_w^2) &= E\left[\sum_{j=1}^{m}\left\{\sum_{i=1}^{\bar{n}}y_{ij}^2 - \bar{n}\,\bar{y}_{.j}^2\right\}\right]\\ &= \sum_{j=1}^{m}\sum_{i=1}^{\bar{n}}Var(y_{ij}) - \bar{n}\sum_{j=1}^{m}Var(\bar{y}_{.j}). \qquad (10.4.10)\end{aligned}$$

Now, taking $\bar{n} = 1$ in (10.4.6), we get

$$Var(y_{ij}) = S_x^2 + S_\epsilon^2 + S_\alpha^2. \qquad (10.4.11)$$

Substitution from (10.4.11) and (10.4.6) in (10.4.10), we obtain

$$E(s_w^2) = S_x^2 + S_\epsilon^2. \qquad (10.4.12)$$

Let s_u^2 denote the mean sum of squares between the means of units defined by

$$s_u^2 = \frac{1}{h-1}\sum_{i=1}^{h}(\bar{y}_{i\cdot} - \bar{y}_{\cdot\cdot})^2.$$

Then, we get

$$(h-1)E(s_u^2) = E\left[\sum_{i=1}^{h}\bar{y}_{i\cdot}^2 - h\bar{y}_{\cdot\cdot}^2\right]$$

$$= \sum_{i=1}^{h} Var(\bar{y}_{i\cdot}) - h\,Var(\bar{y}_{\cdot\cdot}). \qquad (10.4.13)$$

We obtain $Var(\bar{y}_{i\cdot})$ from (10.3.13) by taking $h = 1$, $m = p$, and $n = p$. Hence

$$Var(\bar{y}_{i\cdot}) = S_x^2 + \frac{1}{p}S_\alpha^2 + \frac{1}{p}S_\epsilon^2. \qquad (10.4.14)$$

Substitution from (10.4.14) and (10.3.13) in (10.4.13) gives

$$E(s_u^2) = S_x^2 + \frac{h(m-p)}{mp(h-1)}S_\alpha^2 + \frac{1}{p}S_\epsilon^2. \qquad (10.4.15)$$

The three equations (10.4.8), (10.4.12) and (10.4.15) will provide unbiased estimators of S_x^2, S_α^2 and S_ϵ^2 from which one can obtain an estimator of the variance of $\bar{y}_{\cdot\cdot}$.

We consider a particular case. In practice, $p = 1$. Hence, taking $p = 1$ in (10.4.8), from (10.4.12) and (10.4.15), we obtain

$$E(s_e^2) = \frac{m}{n}(S_x^2 + S_\epsilon^2) + S_\alpha^2 \qquad (10.4.16)$$

$$E(s_w^2) = S_x^2 + S_\epsilon^2 \qquad (10.4.17)$$

$$E(s_u^2) = S_x^2 + \frac{h(m-1)}{m(h-1)}S_\alpha^2 + S_\epsilon^2. \qquad (10.4.18)$$

Note that (10.4.16), (10.4.17) and (10.4.18) are not independent. When $p = 1$, we have the following identity

$$(h-1)s_u^2 = \frac{h(m-1)}{m}s_e^2 + (h-m)s_w^2. \qquad (10.4.19)$$

Using (10.4.17) in (10.4.16), we get an unbiased estimator of S_α^2 as

$$Est.S_\alpha^2 = s_e^2 - \frac{m}{h}s_w^2. \qquad (10.4.20)$$

Using (10.4.17) and (10.4.20) in (10.3.14), we obtain an unbiased estimator of $V(\bar{y}_{..})$ as

$$Est.Var(\bar{y}_{..}) = \frac{s_e^2}{m}. \tag{10.4.21}$$

Using (10.4.19) in (10.4.21), we obtain an alternative expression for the estimator of the variance of $\bar{y}_{..}$ as

$$Est.Var(\bar{y}_{..} = \frac{s_u^2}{h} + \frac{(h-m)}{h(m-1)}(s_u^2 - s_w^2). \tag{10.4.22}$$

From (10.4.22) we see that $\frac{s_u^2}{h}$ is no longer an unbiased estimator of variance of $\bar{y}_{..}$, but it is inflated by $\frac{(h-m)}{h(m-1)}(s_u^2 - s_w^2)$, which vanishes if $s_u^2 = s_w^2$.

10.5 Optimum Value of the Number of Enumerators

From (10.3.13), we see that the variance of the sample mean $\bar{y}_{..}$ decreases as the number of the enumerators increases. However, in practice, there is a limit to the number of enumerators to be employed. We shall consider here the problem of determining the values of h and m which will minimize the variance of $\bar{y}_{..}$ for fixed cost of the survey, C_0 say. We assume the cost function as

$$C = c_1 h + c_2 m + c_3\sqrt{hm}$$

where c_1 is the cost per unit of collecting information, c_2 is the cost of employing an enumerator and c_3 is proportional to the cost of travel per unit distance. Here we have assumed that a sample of h units is selected with SRSWOR and is randomly and equally distributed among the m enumerators.

We shall minimize the variance

$$Var(\bar{y}_{..}) = \frac{S_y^2}{h} + \left(\frac{1}{m} - \frac{1}{h}\right) S_\alpha^2 \tag{10.5.1}$$

subject to the condition

$$C_0 = c_1 h + c_2 m + c_3\sqrt{hm}. \tag{10.5.2}$$

We minimize

$$\phi = \frac{S_y^2}{h} + \left(\frac{1}{m} - \frac{1}{h}\right) S_\alpha^2 + \lambda(c_1 h + c_2 m + c_3\sqrt{hm} - C_0) \tag{10.5.3}$$

with respect to h and m, where λ is the Lagrangian multiplier. Equating the partial derivatives of ϕ w.r.t. h and m to zero, we obtain

$$\frac{\delta\phi}{\delta h} = -\frac{S_y^2}{h^2} + \frac{1}{h^2}S_\alpha^2 + \lambda\left(c_1 + \frac{c_3\sqrt{m}}{2\sqrt{h}}\right) = 0$$

$$\frac{\delta\phi}{\delta m} = -\frac{S_\alpha^2}{m^2} + \lambda\left(c_2 + \frac{c_3\sqrt{h}}{2\sqrt{m}}\right) = 0.$$

Eliminating λ between the above two equations, we obtain

$$x^4 + \frac{c_3}{2c_2}x^3 - \frac{c_3}{2c_1}\beta^2 x - \beta^2 = 0, \tag{10.5.4}$$

where $x^2 = \frac{m}{h}$, $\beta^2 = \frac{c_1}{c_2} \cdot \frac{S_\alpha^2}{S_y^2 - S_\alpha^2}$.

An examination of equation (10.5.4) shows that it has two roots, one positive and one negative. The roots of (10.5.4) are to be determined by trial and error method and the positive root $\hat{x} = \sqrt{\frac{\hat{m}}{n}}$ is to be considered. Then, from (10.5.2), by dividing it by h, we get

$$\hat{h} = \frac{C_0}{c_2\hat{x}^2 + c_3\hat{x} + c_1} \tag{10.5.5}$$

and hence from $\hat{x}^2 = \frac{\hat{m}}{\hat{h}}$, we obtain

$$\hat{m} = \frac{C_0\hat{x}^2}{c_2\hat{x}^2 + c_3\hat{x} + c_1}. \tag{10.5.6}$$

A particular case of interest is when $c_3 = 0$. From equation (10.5.4), we obtain

$$\hat{x}^4 = \frac{\hat{m}^2}{\hat{h}^2} = \beta^2$$

and from (10.5.5) and (10.5.6), by putting $c_3 = 0$ we obtain

$$\hat{h} = \frac{C_0}{c_1 + c_2\beta} \tag{10.5.7}$$

and

$$\hat{m} = \frac{C_0\beta}{c_1 + c_2\beta}. \tag{10.5.8}$$

10.6 Problem of Non-response

It is a common experience in sample surveys that data cannot always be collected for all units selected in the sample. This may occur due to refusal by respondents to give information, or their not being at home, or sample units not being accessible, etc. The error in this case would arise because the set of units getting excluded may have characteristics different from the set of units actually surveyed and thus the results of the survey may be biased. We call this type of error the non-response error. This error particularly arises in mail surveys. Due to some units in the sample getting excluded, we have incomplete coverage of the sample.

We shall now explain the technique developed by Hansen and Hurwitz (1946) to deal with the problem of non-response.

Suppose in a SRSWOR of n units selected from a population of N units, n_1 units respond and $n_2 = n - n_1$ units do not respond, in the first attempt. Let a sub-sample of h_2 units be selected from the n_2 non-responding units with SRSWOR and information is collected from them by making special efforts. Let $n_2 = fh_2$, f being assumed constant. We suppose that the population is divided into two classes, those who will respond at the first attempt and those who will not. Those classes will be called as the response and the non-response classes. Let N_1 and N_2 be the sizes of the response and the non-response classes. We consider n_1 units as a SRSWOR from the response class and n_2 units as a SRSWOR from the non-response class. Let $P = \frac{N_1}{N}$ and $Q = 1 - P = \frac{N_2}{N}$. Then, clearly, we have

$$E(n_1) = np = n \cdot \frac{N_1}{N},$$

$$E(n_2) = n(1 - P) = n\frac{N_2}{N}.$$

Let $\bar{y}_{n1}$ and $\bar{y}_{h2}$ denote respectively the means of n_1 units who responded at the first attempt and h_2 units who respond at the second attempt. Then, we prove the following theorem.

Theorem 10.6.1. *An unbiased estimator of the population mean is given by*

$$\bar{y}_w = \frac{n_1\bar{y}_{n1} + n_2\bar{y}_{h2}}{n}.$$

Proof. We have

$$E\left[\frac{n_1}{n}\bar{y}_{n_1}\right] = E_1\left[E_2\left(\frac{n_1}{n}\bar{y}_{n_1}|n_1\right)\right]$$
$$= E_1\left[\frac{n_1}{n}\bar{Y}_{N_1}\right]$$
$$= \bar{Y}_{N_1}P) \tag{10.6.1}$$

where $\bar{Y}_{N_1}$ is the mean of the response class.

Next

$$E\left[\frac{n_2}{n}\bar{y}_{h_2}\right] = E_1\left[E_2\left\{E_3\left(\frac{\frac{n_2}{n}\bar{y}_{h2}}{h_2}\right)|h_2|n_2\right\}\right]$$
$$= E_1\left[E_2\left\{\frac{n_2}{n}\bar{y}_{n2}|n_2\right\}\right]$$
$$= E_1\left[\frac{n_2}{n}\bar{Y}_{N2}\right]$$
$$= \bar{Y}_{n2}Q, \tag{10.6.2}$$

where $\bar{Y}_{N2}$ is the mean of the non-response class.

Using (10.6.1) and (10.6.2), we get

$$E(\bar{y}_w) = \bar{Y}_{N1}P + \bar{Y}_{N2}Q$$
$$= \frac{N_1\bar{Y}_{N1} + N_2\bar{Y}_{N2}}{N}$$
$$= \bar{Y}_N, \tag{10.6.3}$$

where $\bar{Y}_N$ is the mean of the population.

In the next theorem, we derive the variance of $\bar{y}_w$.

Theorem 10.6.2. *The variance of $\bar{y}_w$ is given by*

$$Var(\bar{y}_w) = \left(\frac{1}{n} - \frac{1}{N}\right)S^2 + \frac{f-1}{n}\cdot\frac{N_2}{N}S_2^2,$$

where S^2 and S_2^2 are the mean sum of squares for the whole population and the non-response class respectively.

Proof. We have

$$Var(\bar{y}_w) = V_1[E_2(\bar{y}_w|n_1, n_2)] + E_1[V_2(\bar{y}_w|n_1, n_2)]$$
$$= V_1(\bar{y}_n) + E_1\left[\frac{n_2^2}{n^2}\left(\frac{1}{h_2} - \frac{1}{n_2}\right)s_2^2\right], \tag{10.6.4}$$

where $\bar{y}_n$ = mean of n units and s_2^2 is the mean sum of squares for the sample of n_2 units. Clearly,

$$V_1(\bar{y}_n) = Var(\bar{y}_n) = \left(\frac{1}{n} - \frac{1}{N}\right) S^2, \tag{10.6.5}$$

where S^2 is the mean sum of squares for the whole population. Now,

$$\begin{aligned} E_1\left[\frac{n_2^2}{n^2}\left(\frac{1}{h_2} - \frac{1}{n_2}\right) s_2^2\right] &= E_1\left[\frac{n_2}{n^2}(f-1)s_2^2\right] \\ &= (f-1)E_1\left[E_2\left(\frac{n_2}{n^2}s_2^2 | n_2\right)\right] \\ &= (f-1)E_1\left[\frac{n_2}{n^2}S_2^2\right] \\ &= \frac{(f-1)}{n^2}S_2^2 E_1(n_2) \\ &= \frac{(f-1)}{n^2}S_2^2 \cdot n \cdot Q \\ &= \frac{(f-1)}{n} \cdot \frac{N_2}{N} S_2^2. \end{aligned} \tag{10.6.6}$$

Using (10.6.5) and (10.6.6), we get

$$Var(\bar{y}_w) = \left(\frac{1}{n} - \frac{1}{N}\right) S^2 + \frac{(f-1)}{n} \cdot \frac{N_2}{N} S_2^2. \tag{10.6.7}$$

We shall now consider the problem of determining the optimum values of n and f. For this purpose, we consider the following cost function

$$c = c_0 n + c_1 n_1 + c_2 h_2,$$

where c_0 is the cost of including a sample unit in the initial sample, c_1 is the cost of collecting and processing the information per unit at the first attempt (i.e. in the response class), and c_2 is the cost collecting and processing the information per unit in the non-response class. This cost will vary from sample to sample. So we consider the expected cost $E(c) = C$. We then have

$$C = c_0 n + n c_1 P + \frac{1}{f} c_2 n Q, \tag{10.6.8}$$

since $E(n_1) = nP$, and $E(h_2) = E\left(\frac{n_2}{f}\right) = \frac{nQ}{f}$.

We shall determine the values of n and f which will minimize C for given variance V_0, an estimate of $Var(\bar{y}_w)$.

Theorem 10.6.3. *The values of n and f which will minimize* $C = c_0 n + c_1 nP + \frac{1}{F} c_2 nQ$ *subject to* $V_0 = Var(\bar{y}_w)$ *are given by*

$$\hat{f} = \sqrt{\frac{c_2 \Delta_1}{S_2^2 \Delta_2}},$$

$$\hat{n} = \frac{\Delta_1 + Q\hat{f}S_2^2}{V_0 + \frac{1}{N}S^2},$$

where $\Delta_1 = S^2 - QS_2^2$, *and* $\Delta_2 = c_0 + Pc_1$.

Proof. We minimize

$$\begin{aligned}\phi &= C + \lambda(Var(\bar{y}_w) - V_0) \\ &= c_0 n + c_1 nP + \frac{1}{f}c_2 nQ + \lambda\left[\left(\frac{1}{n} - \frac{1}{N}\right)S^2 + \frac{(f-1)}{n}QS_2^2\right],\end{aligned} \tag{10.6.9}$$

where λ is a Lagrangian multiplier. Equating the partial derivatives of (10.6.9) w.r.t. n and f to zero, we obtain

$$\frac{\delta\phi}{\delta n} = c_0 + c_1 P + \frac{1}{f}c_2 Q + \lambda\left[-\frac{1}{n^2}S^2 - \frac{(f-1)}{n^2}QS_2^2\right] = 0$$

$$\frac{\delta\phi}{\delta f} = -\frac{1}{f^2}c_2 nQ + \frac{\lambda}{n}QS_2^2 = 0.$$

Eliminating λ between the above two equations, we obtain the optimum value of f as

$$\hat{f} = \sqrt{\frac{c_2 \Delta_1}{S_2^2 \Delta_2}}, \tag{10.6.10}$$

where $\Delta_1 = S^2 - QS_2^2$, and $\Delta_2 = c_0 + c_1 P$. The optimum value of n will be obtained from

$$V_0 = \left(\frac{1}{n} - \frac{1}{N}\right)S^2 + \frac{f-1}{n}QS^2$$

and is given by

$$\hat{n} = \frac{\Delta_1 + Q\hat{f}S_2^2}{V_0 + \frac{1}{N}S^2}. \tag{10.6.11}$$

Remark 1. Let n' be the sample size required to ensure $Var(\bar{y}_w) = V_0$ if there were complete response. Then, from $V_0 = \left(\frac{1}{n'} - \frac{1}{N}\right)S^2$, we obtain

$$n' = \frac{NS^2}{NV_0 + S^2}$$

and the optimum value n can be expressed as

$$\hat{n} = n'\left[1 + (\hat{f} - 1)Q\frac{S_2^2}{S^2}\right].$$

Remark 2. If it is assumed that $S^2 = S_2^2$, then

$$\hat{f} = \sqrt{\frac{c_2 P}{c_0 + c_1 P}},$$

$$\hat{n} = n'[1 + (\hat{f} - 1)Q].$$

EXERCISES

10.1. From (10.4.8), (10.4.12) and (10.4.15), obtain unbiased estimators of S_x^2, S_α^2 and S_ϵ^2, and hence obtain an unbiased estimator of variance of sample mean $\bar{y}_{..}$.

10.2. In a post-census survey, it is required to estimate the non-sampling bias and variance for a certain dichotomous characteristic by carefully surveying a sample of n persons selected with SRSWR. Suppose a persons are reported as 0 both in the census and in the survey, whereas b persons are reported as 0 in the census and 1 in the survey and c persons as 1 in the census and 0 in the survey. Assume that the difference between the responses in the census and the survey are uncorrelated. Estimate the non-sampling bias and the variance of the response differences.

10.3. Given that $p = \frac{N_1}{N} = 0.5$, and $S_2^2 = \frac{4}{5}S^2$, and $c_0 = \$1$, $c_1 = \$4$, $c_2 = \$8$, determine the values of n and f which will minimize the expected cost of the survey for fixed variance $V_0 = \frac{S^2}{100}$ of the sample mean. Also, determine the minimum expected cost.

10.4. In a mail survey, n units were selected with SRSWR and n_1 of them responded. From the $n_2 = n - n_1$ non-responding units, r units were selected with SRSWR and the required information was obtained by personal interview. Suggest a suitable unbiased estimator of the population mean $\bar{Y}$ and obtain its variance.

10.5. In Exercise 10.4, the following information is given:

(i) the non-response rate Q of 0.25.

(ii) the population coefficient of variation of the study variable is 100%.

(iii) the ratio of the variance of the non-responding units to the variance for the whole population is 0.5.
(iv) $c_0 = \$0.15$, $c_1 = \$1.00$, $c_2 = \$4.00$,
(v) cost function is given as

$$C = nc_0 + c_1 nP + \frac{1}{f} c_2 nQ.$$

Determine the optimum values of n and f which would minimize the total cost of the survey such that the relative standard error of the estimator is the same as that of a sample of 100 units selected with SRSWR from the population in the event of complete response. Find, also, the minimum cost of the survey.

10.6. In a complete enumeration survey, it is proposed to survey a population of N families over a period of t distinct time points. The time of survey for each family is chosen at random out of t occasions. At the time of survey, the number of occasions the family would be available for data collection is ascertained for each familly. Assuming that each family is available for the data collection at least one of the t occasions, show that the estimator

$$\hat{Y} = \sum_{i=1}^{t} t_i \sum_{j=1}^{N_i} d_{ij} Y_{ij}$$

is unbiased estimator of the population total, where Y_{ij} is the value reported by the jth unit for data collection on the ith occasion, d_{ij} is 1 or 0 accordingly as that unit is contacted on the randomly selected occasion or not, and N_i is the number of units available for data collection on i^{th} occasion.

10.7. Suppose that the data on births in a region are independently collected by the registrar (R) through the method of registration and by an interviewer (I) through the method of enquiry from the households. Let there be

(i) C births recorded by both R and I;
(ii) N_1 births recorded by R, but not by I and found to be correct on verification;
(iii) N_2 births recorded by I, but not by R, and found to be correct on verification; and
(iv) W births recorded by R or I but not by both, and found to be incorrect after verification.

Assuming complete independence between R and I in reporting or missing births, show that an unbiased estimator of N, the total number of births when N is large is given by

$$\hat{N} = C + N_1 + N_2 + \left(\frac{N_1 N_2}{C}\right),$$

and that its variance is approximately equal to $\frac{Nq_1q_2}{p_1p_2}$, where p_1 and p_2 are the probabilitiees of R and I recording a correct birth and $q_1 = 1 - p_1$, $q_2 = 1 - p_2$.

Appendix

Table of Random Numbers

Row No.	1	2	3	4	5	6	7	8	9	10
1	3436	6833	5809	9169	5081	5655	6567	8793	6830	1332
2	6133	4454	2675	3558	7624	5736	2187	4557	0496	8547
3	9853	3890	5535	3045	9830	5455	8218	9090	7266	4784
4	5807	5692	6971	6162	6751	5001	5533	2386	0004	2855
5	6291	0924	1298	7386	5856	2167	8299	9314	0333	8803
6	4725	9516	8555	0379	7746	9647	2010	0979	7155	6653
7	7697	6486	3720	6191	3552	1081	6141	7613	5455	3731
8	3497	2271	9641	0304	4425	6776	1205	2953	5669	1056
9	8940	4765	1641	0606	4970	7582	7991	6480	2946	5190
10	1122	6364	5264	1267	4027	4749	0338	8406	1213	5355
11	4333	0625	3947	1373	6372	9036	7046	4325	3491	8989
12	7685	1550	0853	4276	1572	9348	6893	2113	8285	9195
13	0592	8341	4430	0496	9613	2643	6442	0870	5449	8560
14	3506	0774	0447	7461	4459	0866	1698	0184	4975	5447
15	8368	2507	3565	4243	6667	8324	3063	8809	4248	1190
16	2630	1112	6680	4863	6813	4149	8325	2271	1963	9569
17	8883	3897	1848	8150	8184	1133	6088	3641	6785	0658
18	1123	3943	5248	0635	9265	4052	1509	1280	0953	9107
19	1167	9827	4101	4496	1254	6814	2479	5924	5071	1244
20	7831	0877	3806	9734	3801	1651	7169	3974	1725	9709
21	2487	9756	9886	6776	9426	0820	3741	5427	5293	3223
22	1245	3875	9816	8400	2938	2530	0158	5267	4639	5428
23	5309	4806	3176	8397	5758	2503	1567	5740	2577	8899
24	7109	0702	4179	0438	5234	9480	9777	2858	4391	0979
25	8716	7177	3386	7643	6555	8665	0768	4409	3647	9286
26	9499	5280	5150	2724	8482	8362	1566	2469	9704	8165
27	3125	4552	6044	0222	7520	1521	8205	0599	5167	1654
28	3788	6257	0632	0693	2263	5290	0511	0229	5951	6808
29	2242	2143	8724	1212	9485	3985	7280	0130	7791	6272
30	0900	4364	6429	8573	9904	2269	6405	9459	3088	6903

Row No.	Column Number 1	2	3	4	5	6	7	8	9	10
31	7909	4528	8772	1876	2113	4781	8678	4873	2061	1835
32	0379	2073	2680	8258	6275	7149	6858	4578	5932	9582
33	0780	6661	0277	0998	0432	8941	8946	9784	6693	2491
34	8478	8093	6990	2417	0290	5771	1304	3306	8825	5937
35	2519	7869	9035	4282	0307	7516	2340	1190	8440	6551
36	2472	0823	6188	3303	0490	9486	2896	0821	5999	3697
37	8418	5411	9245	0857	3059	6689	6523	8386	6674	7081
38	8293	5709	4120	5530	8864	0511	5593	1633	4788	1001
39	9260	1416	2171	0525	6016	9430	2828	6877	2570	4049
40	6568	1568	4160	0429	3488	3741	3311	3733	7882	6985
41	6694	5994	7517	1339	6812	4139	6938	8098	6140	2013
42	2273	6882	2673	6903	4044	3064	6738	7554	7734	7899
43	6364	5762	0322	2592	3452	9002	0264	6009	1311	5873
44	6696	1759	0563	8104	5055	4078	2516	1631	5859	1331
45	3431	2522	2206	3938	7860	1886	1229	7734	3283	8487
46	4842	3765	3484	2337	0587	9885	8568	3162	3028	7091
47	8295	9315	5892	6981	4141	1606	1411	3196	9428	3300
48	4925	4677	8547	5258	7274	2471	4559	6581	8232	7405
49	5439	0994	3794	8444	1043	4629	5975	3340	3793	6060
50	2031	0283	3320	1595	7953	2695	0399	9799	6114	2091

References

Journal Abbreviations

AJS: Australian Journal of Statistics

AISM: Annals of the Institute of Statistical Mathematics

AMS: Annals of Mathematical Statistics

BCSA: Bulletin of Calcutta Statistical Association

BISI: Bulletin of the International Statistical Institute

JASA: Journal of the American Statistical Association

JISA: Journal of the Indian Statistical Association

JISAS: Journal of the Indian Society of Agricultural Statistics

JRSS: Journal of the Royal Statistical Society

RISI: Review of the International Statistical Institute

Aoyana, H. (1951). On practical systematic sampling. *AISM* **3**, 57-64.

Armitagge, P. (1947). A comparison of stratified with unrestricted random sampling from a finite population. *Biometrika* **34**, 273-280.

Avadhani, M. S. and Sukhatme, B. V. (1965). Controlled simple random sampling. *JISAS* **17**, 34-42.

Avadhani, M.S. and Sukhatme, B. V. (1973). Controlled sampling with equal probabilities and without replacement. *ISR* **41**, 175-183.

Banerjee, K. S. (1955). A note on successive sampling. *BCSA* **6**, 35-39.

Bartholomew, D. J. (1961). A method of allowing for 'not-at-home' bias in sample surveys. *App. Stat.* **10**, 52-59.

Bartlett, M. S. (1937). Sub-sampling for attributes. *JRSS Supplement* **4**, 131-135.

Basu, D. (1958). On sampling with and without replacement. *Sankhyā* **20**, 287-294.

Bellhouse, D. R. and Rao, J. N. K. (1975). Systematic sampling in the presence of a trend. *Biometrika* **62**, 694-697.

Bose, C. (1943). Note on the sampling error in the method of double sampling. *Sankhyā* **6**, 329-330.

Bowley, A. L. (1926). Measurement of precision attained in sampling. *BISI* **22**, 1-62.

Brewer, K. W. R. (1963a). A model of systematic sampling with unequal probabilities. *AJS* **5**, 5-13.

Brewer, K. W. R. (1963b). Ratio estimation in finite populations: Some results deducible from the assumption of an underlying stochastic process. *AJS* **5**, 93-105.

Brooks, S. (1955). The estimation of an optimum subsampling number. *JASA* **50**, 398-415.

Buckland, W. R. (1951). A review of the literature of systematic sampling. *JRSS* **B13**, 208-215.

Carver, H. C. (1930). Fundamentals of the theory of sampling. *AMS* **1**, 101-121, 260-274.

Chikkagouder, M. S. (1966a). A note on inverse sampling with equal probabilities. *Sankhyā* **28(A)**, 93-96.

Chikkagouder, M. S. (1966b). A note on sampling with varying probabilities. *JISAS* **18**, 86-92.

Cochran, W. G. (1942). Sampling theory when the sampling units are of unequal sizes. *JASA* **37**, 199-212.

Cochran, W. G. (1946). Relative accuracy of systematic and stratitified random samples for a certain class of populations. *AMS* **17**, 164-177.

Cochran, W. G. (1977). *Sampling Techniques*, 3rd ed. John Wiley & Sons, Inc., New York.

Cochran, W. G., Mosteller, F., and Tukey, J. W. (1954). Principles of sampling. *JASA* **49**, 13-35.

Cox, D. R. (1952). Estimation by double sampling. *Biometrika* **39**, 217-227.

Dalenius, T. (1950). The problem of optimum stratification - I. *Skand. Akt.* **33**, 203-213.

Dalenius, T. (1962). Recent advances in sample survey theory and methods. *AMS* **33**, 325-349.

Dalenius, T. and Gurney, M. (1951). The problem of optiimum statification - II. *Skand. Akt.* **34**, 133-148.

Dalenius, T. and Hodges, J. L., Jr. (1959). Minimum variance stratification. *JASA* **54**, 88-101.

Das, A. C. (1950). Two-dimensional systematic sampling and the associated stratified and random sampling. *Sankhyā* **10**, 95-108.

Das, A. C. (1951a). On two phase sampling and samplinlg witih varying probabilities. *BISI* **33(2)**, 105-112.

Das, A. C. (1951b). Systematic sampling. *BISI* **33(2)**, 119-132.

Das, A. C. (1962). On MVU estimates of parameters of a finite population based on varying probability samples. *BCSA* **11**, 39-48.

David, I. P. and Sukhatme, B. V. (1974). On the bias and mean square error of the ratio estimator. *JASA* **69**, 464-466.

Deming, W. E. and Glasser, G. J. (1959). On the problem of matching lists by samples. *JASA* **54**, 403-415.

Des Raj (1954a). On sampling with probabilities proportional to size. *Ganita* **5**, 175-182.

Des Raj (1954b). Ratio estimation in sampling with equal and unequal probabilities. *JISAS* **6**, 127-138.

Des Raj (1956a). A note on the determination of optimum probabilities in sampling without replacement. *Sankhyā* **17**, 197-200.

Des Raj (1956b). Some estimators in sampling with varying probabilities without replacement. *JASA* **51**, 269-284.

Des Raj (1962). On matching lists by samples. *JASA* **56**, 251-255.

Des Raj (1964a). On double sampling for *pps* estimation. *AMS* **35**, 900-902.

Des Raj (1964b). The use of systematic sampling with probability proportional to size in a large-scale survey. *JASA* **59**, 251-255.

Des Raj (1964c). A note on the variance of the ratio estimate. *JASA* **59**, 895-898.

Des Raj (1965). On sampling over two occasions with probability proportional to size. *AMS* **36**, 327-330.

Des Raj (1966). Some remarks on a simple procedure of sampling without replacement. *JASA* **61**, 391-396.

Des Raj (1968). *Sampling Theory.* McGraw-Hill Book Company, New York.

Des Raj and Khamis, S. H. (1958). Some remarks on sampling with replacement. *AMS* **29**, 550-557.

Durbin, J. (1953). Some results in sampling theory when the units are selected with unequal probabilities. *JRSS* **B15**, 262-269.

Durbin, J. (1959). A note on the application of Quenouille's method of bias reduction to the estimation of ratios. *Biometrika*, **46**, 477-480.

Eckler, A. R. (1955). Rotation sampling. *AMS* **26**, 664-685.

Evans, D. H. (1963). Multiplex sampling. *AMS* **34**, 1322-1346.

Evans, W. D. (1951). On stratification and optimum allocation. *JASA* **46**, 95-104.

Fellegi, I. P. (1963). Sampling with varying probabilities without replacement-rotating and non-rotating samples. *JASA* **58**, 183-201.

Fellegi, I. P. (1964). Response variance and its estimation. *JASA* **59**, 1016-1041.

Fisher, R. A. and Yates, F. (1943). *Statistical Tables for Biological, Agricultural and Medical Research*, 2nd ed. Oliver and Boyd Ltd., London.

Foreman, E. K. and Brewer, K. W. R. (1971). The efficient use of supplementary information in standard sampling procedures. *JRSS* **B33**, 391-400.

Gautschi, W. (1957). Some remarks on systematic sampling. *AMA* **28**, 385-394.

Ghosh, B. (1947). Double sampling with many auxiliary variates. *BCSA* **1**, 91-93.

Ghosh, B. (1949). Interpenetrating (networks of) samples. *BCIA* **2**, 108-119.

Ghosh, S. P. (1963). Estimating the mean by two-stage sampling with replacement. *BCSA* **12**, 97-103.

Glasser, G. J. (1962). On estimators of variances and covariances. *Biometrika* **49**, 259-262.

Godambe, V. P. (1951). On two-stage sampling. *JRSS* **B13**, 216-218.

Godambe, V. P. (1955). A unified theory of sampling from finite population. *JRSS* **B17**, 269-278.

Godambe, V. P. (1960). An admissible estimate for any sampling design. *Sankhyā* **22**, 285-288.

Godambe, V. P. (1965). A review of the contributions towards a unified theory of sampling from finite populations. *RISI* **33(2)**, 242-258.

Godambe, V. P. (1966a). A new approach to sampling from finite populations I. *JRSS* **B28**, 310-319.

Godambe, V. P. (1966b). A new approach to sampling from finite populations II. *JRSS* **B28**, 320-328.

Goodman, L. A. (1960). On the exact variance of products. *JASA* **55**, 708-713.

Goodman, L. A. and Hartley, H. O. (1958). The precision of unbiased ratio-type estimators. *JASA* **53**, 491-508.

Hajek, J. (1958). Some contributions to the theory of probability sampling. *BISI* **36(3)**, 127-133.

Hajek, J. (1960). On the theory of ratio estimates. *BISI* **37(2)**, 219-226.

Hannan, E. J. (1962). Systematic sampling. *Biometrika* **49**, 281-283.

Hansen, M. H. and Hurwitz, W. N. (1942). Relative efficiencies of various sampling units in population inquiries. *JASA* **37**, 89-94.

Hansen, M. H. and Hurwitz, W. N. (1943). On the theory of sampling from finite populations. *AMS* **14**, 333-362.

Hansen, M. H. and Hurwitz, W. N. (1946). The problem of non-response in sample surveys. *JASA* **41**, 517-529.

Hansen, M. H., Hurwitz, W. N., and Bershad, M. (1961). Measurement errors in censuses and surveys. *BISI* **38(2)**, 359-374.

Hansen, M. H., Hurwitz, W. N., and Madow, W. G. (1953). *Sample Survey Methods and Theory.* John Wiley & Sons, New York. Vols. I and II.

Hanurao, T. V. (1962a). Some sampling schemes in probabiilty sampling. *Sankhyā* **24(A)**, 421-428.

Hanurao, T. V. (1962b). On Horvitz and Thompson estimator. *Sankhyā* **24(A)**, 429-436.

Hanurao, T. V. (1966). Some aspects of unified sampling theory. *Sankhyā* **28(A)**, 175-203.

Hartley, H. O. (1966). Systematic sampling with unequal probability and without replacement. *JASA* **61**, 739-748.

Hartley, H. O. and Rao, J. N. K. (1962). Sampling with unequal probabilities and without replacement. *AMS* **33**, 350-374.

Hartley, H. O. and Rao, J. N. K. (1968). A new estimation theory for sample surveys. *Biometrika* **55**, 547-557.

Hartley, H. O., Rao, J. N. K., anad Kiefer, G. (1969). Variance estimation with one unit per stratum. *JASA* **64**, 841-851.

Hartley, H. O. and Ross, A. (1954). Unbiased ratio estimates. *Nature* **174**, 270-271.

Hendricks, W. A. (1956). *The Mathematical Theory of Sampling.* Scarecrow Press, New Brunswick, N. J.

Horvitz, D. G., Greenberg, B. G., and Abernathy, J. R. (1975). Some recent developments in randomized response designs. *A Survey of Statistical Design and Linear Models.* J. N. Srivastava (ed.). American Elsevier Publishing Co., New York, 271-285.

Horvitz, D. G. and Thompson, D. J. (1952). A generalization of sampling without replacement from a finite universe. *JASA* **47**, 663-685.

Hege, V. S. (1965). Sampling designs which admit uniformly minimum variance unbiased estimates. *BCSA* **14**, 160-162.

Jessen, R. J. (1942). Statistical investigation of a sample survey for obtaining farm facts. *Iowa Agricultural Experimental Station Research Bulletin*, No. 304.

Joshi, V. M. (1965a). Admissibility and Baye's estimation on sampling from fininte populations - II. *AMS* **36**, 1723-1729.

Joshi, V. M. (1965b). Admissibility and Baye's estimation in sampling finite populations - III. *AMS* **36**, 1730-1742.

Joshi, V. M. (1966). Admissibility and Baye's estimation in sampling finite populations - IV. *AMS* **37**, 1658-1670.

Kempthorne, O. (1969). Some remarks on inference in fininte sampling. *New Developments in Survey Sampling.* N. L. Johnson and H. Smith, Jr. (ed.) John Wiley & Sons, New York, 671-695.

Kendall, M. G. and Smith, B. B. (1938). Randomness and random sampling numbers. *JRSS* **(A)101**, 147-166.

Keyfitz, N. (1957). Estimates of sampling variance when two units are selected from each stratum. *JASA* **52**, 503-510.

Khan, S. and Tripathi, T. P. (1967). The use of multivariate auxiliary information in double sampling. *JISA* **5**, 42-48.

Kitagawa, T. (1955). Some contributions to the design of sample surveys, Part I, II, & III. *Sankhyā* **14**, 317-362.

Kitagawa, T. (1956). Some contributions to the design of sample surveys, Parts IV, V, & VI. *Sankhyā* **17**, 1-36.

Koop, J. C. (1951). A note on the bias of the ratio eestimator. *BISI* **33(1)**, 141-146.

Koope, J. C. (1968). An exercise in ratio estimation. *Amer. Statistician* **22**, 29-30.

Kulldorf, G. (1963). Some problems of optimum allocation for sampling on two occasions. *RISI* **31**, 24-57.

Lahiri, D. B. (1951). A method of sample selection providing unbiased ratio estimates. *BISI* **33**, 133-140.

Lieberman, G. J. and Owen, D. B. (1961). *Tables of the Hypergeometric Probability Distribution.* Stanford University Press, Stanford, California.

Madhava, K. B. (1939). Techniques of random sampling. *Sankhyā* **4**, 532-534.

Madow, L. H. (1946). Systematic sampling and its relation to other sampling designs. *JASA* **41**, 204-217.

Madow, W. G. (1949). On the theory of systematic sampling II. *AMS* **20**, 333-354.

Madow, W. G. (1953). On the theory of systematic sampling III. *AMS* **24**, 101-106.

Madow, W. G. and Madow, L. H. (1944). On the theory of systematic sampling I. *AMS* **15**, 1-24.

Mahalanobis, P. C. (1946). Recent experiments in statistical sampling in the Indian Statistical Institute. *JRSS* **(A)109**, 325-370.

Mathai, A. (1954). On selecting random numbers for large-scale sampling. *Sankhyā* **13**, 157-160.

Mickey, M. R. (1959). Some finite population unbiased ratio and regression estimators. *JASA* **54**, 594-612.

Midzuno, H. (1951). On the sampling system with probability proportionate to sum of sizes. *AISM* **2**, 99-108.

Milne, A. (1959). The centric systematic area sample treated as a random sample. *Biometrics* **15**, 270-297.

Mokashi, V. K. (1950). A note on interpenetrating samples. *JISAS* **2**, 189-195.

Murthy, M. N. (1957). Ordered and unordered estimators in sampling without replacement. *Sankhyā* **18**, 379-390.

Murthy, M. N. (1963a). Some recent advances in sampling theory. *JASA* **58**, 737-755.

Murthy, M. N. (1963b). A note on determination of sample size. *Sankhyā* **25(A)**, 381-382.

Murthy, M. N. (1963c). Generalized unbiased estimation in sampling from finite populations. *Sankhyā* **25(B)**, 245-262.

Murthy, M. N. (1964). Product method of estimation. *Sankhyā* **26(A)**, 69-74.

Murthy, M. N. (1967). *Sampling Theory and Methods.* Statistical Publishing Society, Calcutta, India.

Murthy, M. N. and Nanjamma, N. S. (1959). Almost unbiased ratio estimates based on interpenetrating sub-sample estimates. *Sankhyā* **21**, 381-392.

Nanjamma, N. S., Murthy, M. N., and Sethi, V. K. (1959). Some sampling systems providing unbiased ratio estimates. *Sankhyā* **21**, 299-314.

Narain, R. D. (1951). On sampling without replacement with varying probabilities. *JISAS* **5**, 96-99.

Nordin, J. A. (1944). Determining sample size. *JASA* **39**, 497-506.

Olkin, I. (1958). Multivariate ratio estimation for finite populations. *Biometrika* **45**, 154-165.

Pasoual, J. N. (1961). Unbiased ratio estimators in stratified sampling. *JASA* **56**, 70-87.

Pathak, P. K. (1961). Use of order statistic in sampling without replacement. *Sankhyā* **23(A)**, 409-414.

Pathak, P. K. (1962a). On simple random sampling with replacement. *Sankhyā* **24(A)**, 287-302.

Pathak, P. K. (1962b). On sampling units with unequal probabilities. *Sankhyā* **24(A)**, 315-326.

Pathak, P. K. (1964a). On sampling schemes providing unbiased ratio estimates. *AMS* **35**, 222-231.

Pathak, P. K. (1964b). On inverse sampling with unequal probabilities. *Biometrika* **51**, 185-193.

Pathak, P. K. and Shukla, N. D. (1966). Non-negativity of a variance estimator. *Sankhyā* **28(A)**, 41-46.

Patterson, H. D. (1950). Sampling on successive occasions with partial replacement of units. *JRSS* **B12**, 241-255.

Patterson, H. D. (1954). The errors of lattice sampling. *JRSS* **B16**, 140-149.

Prabhu Ajgaonkar, S. G. (1965). On a class of linear estimators in sampling with varying probabilities without replacement. *JASA* **60**, 637-642.

Quenouille, M. H. (1956). Notes on bias in estimation. *Biometrika* **43**, 353-360.

Rand Corporation (1955). *A Million Random Digits.* Free Press, Glencoe, IL.

Rao, C. R. (1971). Some aspects of statistical inference in problems of sampling from finite populations. *Foundations of Statistical Inference.* V. P. Godambe and D. A. Sprott (eds.) Holt, Rinehart, and Winston, Toronto, Canada, 177-202.

Rao, C. R., Mathai, A., and Mitra, S. K. (1966). *Formulae and Tables for Statistical Work.* Statistical Publishing Society, Calcutta, India.

Rao, J. N. K. (1962). On the estimation of relative efficiency of sampling procedures. *AISM* **14**, 143-150.

Rao, J. N. K. (1965). On two simple schemes of unequal probability sampling without replacement. *JISA* **3**, 173-180.

Rao, J. N. K. (1968). Some small sample results in ratio and regression estimation. *JISA* **6**, 160-168.

Rao, J. N. K. (1969). Ratio and regression estimators. *New Developments in Survey Sampling.* N. L. Johnson and H. Smith, Jr. (eds.) John Wiley & Sons, New York, 213-234.

Rao, J. N. K. (1975). On the foundations of survey sampling. *A Survey of Statistical Design and Linear Models.* J. N. Srivastava (ed.) American Elsevier Publishing Co., New York, 489-505.

Rao, J. N. K., Hartley, H. O., and Cochran, W. G. (1962). A simple procedure of unequal probability sampling without replacement. *JRSS* **B24**, 482-491.

Rao, J. N. K. and Pereira, N. P. (1968). On double ratio estimators. *Sankyā* **A30**, 83-90.

Rao, T. J. (1966a). On certain unbiased ratio estimators. *AISM*, **18**, 117-121.

Rao, T. J. (1966b). On the variance of the ratio estimator for Midzuno-Sen sampling scheme. *Metrika* **10**, 89-91.

Rao, P. S. R. S. and Mudholkar, G. S. (1967). Generalized multivariate estimators for the mean of finite populations. *JASA* **62**, 1009-1012.

Robson, D. S. (1957). Applications of multivariate polykays to the theory of unbiased ratio type estimation. *JASA* **52**, 511-522.

Roy, J. and Chakravarti, I. M. (1960). Estimating the mean of finite population. *AMS* **31**, 392-398.

Royall, R. M. (1968). An old approach to finite population sampling theory. *JASA* **63**, 1269-1279.

Royall, R. M. (1970a). On finite population sampling theory under certain linear regression models. *Biometrika* **57**, 377-387.

Royall, R. M. (1970b). Finite population sampling-on labels in estimation. *AMS* **41**, 1774-1779.

Royall, R. M. (1971). Linear regression models in finite population sampling theory. *Foundations of Statistical Inference.* V. P. Godambe and D. A. Sprott (eds.) Holt, Rinehart, & Winston, Toronto, Canada, 259-279.

Sampford, M. R. (1967). On sampling without replacement with unequal probabilities of selection. *Biometrika* **54**, 499-513.

Sen, A. R. (1953a). Recent advances in sampling with varying probabilities. *BCSA* **5**, 1-15.

Sen, A. R. (1953b). On the estimate of the variance in sampling with varying probabilities. *JISAS* **5**, 119-127.

Sen, A. R. (1972). Successive sampling with p ($p \geq 1$) auxiliary variables. *AMS* **43**, 2031-2034.

Sen, A. R. (1973a). Theory and application of sampling on repeated occasions with several auxiliary variables. *Biometrics* **29**, 383-385.

Sen, A. R. (1973b). Some theory of sampling on successive occasions. *AJS* **15**, 105-110.

Seth, G. R. anad Rao, J. N. K. (1964). On the comparison between simple random sampling with and without replacement. *Sankhyā* **A26**, 85-86.

Singh, D., Jindal, K. K., and Gary, J. N. (1968). On modified systematic sampling. *Biometrika* **55**, 541-546.

Smith, H. F. (1938). An empirical law describing heterogeneity in the yields of agricultural crops. *J. Agri. Soc.* **28**, 1-23.

Smith, T. M. F. (1976). The foundations of survey sampling. A review. *JRSS* **A139**, 183-204.

Stein, C. (1945). A two-sample test for a linear hypothesis whose power is independent of the variance. *AMS* **16**, 243-258.

Sukhatme, P. V. and Seth, G. R. (1952). Non-sampling errors in surveys. *JISAS*, **4**, 5-41.

Sukhame, P. V. and Sukhatme, B. V. (1970). *Sampling Theory of Surveys with Applications.* Iowa State University Press, Ames, Iowa, USA.

Tin, M. (1965). Comparison of some ratio estimators. *JASA* **60**, 294-307.

Tippett, L. H. C. (1927). *Random Sampling Numbers.* CUP, London.

Tukey, J. W. (1950). Some sampling simplified. *JASA* **45**, 501-519.

Watson, D. J. (1937). The estimation of leaf areas. *J. Agri. Sci.* **27**, 474.

Williams, W. H. (1963). The precision of some unbiased regression estimators. *Biometrika* **17**, 267-274.

Wishart, J. (1952). Moment-coefficients of the k-statistics in samples from a finite population. *Biometrika* **39**, 1-13.

Yates, F. (1948). Systematic sampling. *Phil. Trans. Roy. Soc. London* **A241**, 345-377.

Yates, F. (1960). *Sampling Methods for Censuses and Surveys*, 3rd edition. Charles Griffin and Co., London.

Yates, F. and Grundy, P. M. (1953). Selection without strata with probability proportional to size. *JRSS* **B15**, 253-261.

Zarkovic, S. S. (1960). On the efficiency of sampling with various probabilities and the selection of units with replacement. *Metrika* **3**, 53-60.

Index